Lecture Notes in Computer Science 13392

More information about this series at https://link.springer.com/bookseries/558

Robert Krimmer · Marius Rohde Johannessen ·
Thomas Lampoltshammer · Ida Lindgren ·
Peter Parycek · Gerhard Schwabe ·
Jolien Ubacht (Eds.)

Electronic Participation

14th IFIP WG 8.5 International Conference, ePart 2022
Linköping, Sweden, September 6–8, 2022
Proceedings

Editors
Robert Krimmer
University of Tartu
Tartu, Estonia

Thomas Lampoltshammer
Danube University Krems
Krems, Austria

Peter Parycek
Danube University Krems
Krems, Austria

Jolien Ubacht
Delft University of Technology
Delft, The Netherlands

Marius Rohde Johannessen
University of South-Eastern Norway
Borre, Norway

Ida Lindgren
Linköping University
Linköping, Sweden

Gerhard Schwabe
University of Zurich
Zurich, Switzerland

ISSN 0302-9743 ISSN 1611-3349 (electronic)
Lecture Notes in Computer Science
ISBN 978-3-031-23212-1 ISBN 978-3-031-23213-8 (eBook)
https://doi.org/10.1007/978-3-031-23213-8

This Springer imprint is published by the registered company Springer Nature Switzerland AG
The registered company address is: Gewerbestrasse 11, 6330 Cham, Switzerland

Preface

The EGOV-CeDEM-ePart 2022 conference, or for short EGOV2022, is now in the fifth year of its existence after the successful merger of three formerly independent conferences, i.e., the IFIP WG 8.5 International Conference on Electronic Government (EGOV), the Conference for E-Democracy and Open Government Conference (CeDEM), and the IFIP WG 8.5 International Conference on Electronic Participation (ePart). This larger, united conference is dedicated to a broad area of digital or electronic government, open government, smart governance, artificial intelligence, e-democracy, policy informatics, and electronic participation. Scholars from around the world have found this conference to be a premier academic forum with a long tradition along its various branches, which has given the EGOV-CeDEM-ePart conference its reputation of the leading conference worldwide in the research domains of digital/electronic, open, and smart government as well as electronic participation.

The call for papers attracted completed research papers, work-in-progress papers on ongoing research (including doctoral papers), project and case descriptions, and workshop and panel proposals. This volume contains only full papers. All submissions were assessed through a double-blind peer-review process, with at least three reviewers per submission, and the acceptance rate was 46%. The acceptance rate was higher than in the previous years as there were many high-quality papers. The review time took 39 days this year, thanks to the contributions of the many Program Committee (PC) members.

The review process was focused on ensuring double-blind reviewing and avoiding any conflicts of interest. Authors submitted their papers to a particular track. The track chairs handled the papers within their own track by assigning reviewers and proposing acceptance decisions. The lead track chair became part of the editor of the proceedings, in addition to the general chairs. Track chairs were not allowed to submit to their own track, nor were persons from the same university or close collaborators of a track chair to avoid any conflict of interest. Track chairs could either submit to another track or to the 'track chairs' track. The latter was handled by the general chairs. The general chairs checked if there was any conflict of interest among the papers submitted to each track, if so, then papers were moved to another track. The track chairs checked that all papers were submitted anonymously. If not, the authors were asked to resubmit within days. Track chairs assigned the reviewers and selected the Programme Committee members in such a way that there were no conflicts of interest. After at least three reviews were received, the track chairs made a proposal for a decision per paper. The decisions were discussed in a meeting with the general and track chairs to ensure that the decisions were made in a consistent manner per track.

The conference tracks of the 2022 edition present the evolution of the topics and the progress in this field. The papers were distributed over the following tracks:

- General E-Government and E-Governance
- General E-Democracy and E-Participation
- ICT and Sustainability Development Goals

- AI, Data Analytics, and Automated Decision Making
- Digital and Social Media
- Digital Society
- Emerging Issues and Innovations
- Legal Informatics
- Open Data
- Smart and Digital Cities (Government, Districts, Communities and Regions)

Among the full research paper submissions, 12 papers (empirical and conceptual) were accepted for this year's Springer LNCS ePart proceedings (vol. 13392) from the General E-Democracy and E-Participation, ICT and Sustainability, Digital and Social Media, Legal Informatics, and Digital Society tracks.

The Springer LNCS EGOV proceedings (vol. 13391) contain the completed research papers from the General E-Government and E-Governance; AI, Data Analytics, and Automated Decision Making; Emerging Issues and Innovations; Open Data; and Smart and Digital Cities tracks.

The papers included in this volume have been clustered under the following headings:

- E-Democracy and E-Participation
- ICT and Sustainability
- Digital and Social Media
- Legal Informatics
- Digital Society

As in previous years and per the recommendation of the Paper Awards Committee under the leadership of Noella Edelmann from Danube University Krems, Austria, and Evangelos Kalampokis from the University of Macedonia, Greece, the IFIP EGOV-CeDEM-ePart 2022 Conference Organizing Committee granted outstanding paper awards in three distinct categories:

- The most interdisciplinary and innovative research contribution
- The most compelling critical research reflection
- The most promising practical concept

The winners in each category were announced during the obligatory awards ceremony at the conference.

Many people behind the scenes make large events like this conference happen. We would like to thank the members of the Program Committee, the reviewers, and the track chairs for their great efforts in reviewing the submitted papers. We would also like to express our deep gratitude to Ulf Melin and the local organization team for hosting the conference.

The EGOV-CeDEM-ePart 2022 conference was hosted by the Division of Information Systems and Digitalization at the Department of Management and Engineering, Linköpings Universitet (LiU). LiU conducts world-leading, boundary-crossing research in fields including materials science, IT, and hearing. In the same spirit, LiU offers many innovative educational programs, many of them with a clear vocational focus, leading to qualification as, for example, doctors, teachers, economists and engineers. LiU has 35,900 students and 4,300 employees on four campuses. The conference was held at the largest campus - Campus Valla – which is situated just outside the city center of Linköping. We were very happy to be hosted here and enjoyed the newly opened building, having many in-depth discussions advancing the EGOV-CeDEM-ePart field.

We hope that the papers help to advance your research and hope that you will enjoy reading them.

September 2022

Robert Krimmer
Marius Rohde Johannessen
Thomas Lampoltshammer
Ida Lindgren
Peter Parycek
Gerhard Schwabe
Jolien Ubacht

Organization

Conference and Program Committee Chairs

Robert Krimmer	University of Tartu, Estonia
Marius Rohde Johannessen	University of South-Eastern Norway, Norway
Thomas Lampoltshammer	Danube University Krems, Austria
Ida Lindgren	Linköping University, Sweden
Peter Parycek	Danube University Krems, Austria
Gerhard Schwabe	University of Zurich, Switzerland
Jolien Ubacht	Delft University of Technology, The Netherlands

Program Committee

Karin Ahlin	Mid Sweden University, Sweden
Salah Uddin Ahmed	University of South-Eastern Norway, Norway
Suha Alawadhi	Kuwait University, Kuwait
Valerie Albrecht	Danube University Krems, Austria,
Cristina Alcaide Muñoz	University of Alcalá, Spain
Laura Alcaide-Muñoz	University of Granada, Spain
Konstantina Alexouda	International Hellenic University, Greece
Leonidas Anthopoulos	University of Thessaly, Greece
Ari-Veikko Anttiroiko	Tampere University, Finland
Wagner Araujo	United Nations University, Portugal
Karin Axelsson	Linköping University, Sweden
Luiza Azambuja	Tallinn University of Technology, Estonia
Dian Balta	Fortiss, Germany
Kristina Belancic	Danube University Krems, Austria
Peter Bellström	Karlstad University, Sweden
Lasse Berntzen	University of South-Eastern Norway, Norway
Christina Bidmon	Utrecht University, The Netherlands
Radomir Bolgov	Saint Petersburg State University, Russia
Alessio Maria Braccini	University of Tuscia, Italy
Paul Brous	Delft University of Technology, The Netherlands
Matthias Buchinger	Technische Universität München, Germany
Kelvin Joseph Bwalya	University of Johannesburg, South Africa
Jesus Cano,	National University of Distance Education (UNED), Spain
Iván Cantador	Universidad Autónoma de Madrid, Spain

João Carvalho	University of Minho, Portugal
Luiz Paulo Carvalho	Federal University of Rio de Janeiro, Brazil
Youngseok Choi	University of Southampton, UK
Wichian Chutimaskul	King Mongkut's University of Technology Thonburi, Thailand
Vincenzo Ciancia	Istituto di Scienza e Tecnologie dell'Informazione "A. Faedo", Consiglio Nazionale delle Ricerche, Italy
Antoine Clarinval	Université de Namur, Belgium
Taiane Coelho	Federal University of Parana, Brazil
Andreiwid Sheffer Corrêa	Federal Institute of Sao Paulo, Brazil
María E Cortés Cediel	Universidad Complutense de Madrid, Spain
J. Ignacio Criado	Universidad Autónoma de Madrid, Spain
Joep Crompvoets	KU Leuven, Belgium
Peter Cruickshank	Edinburgh Napier University, UK
Jonathan Crusoe	Gothenburg University and University of Borås, Sweden
Csaba Csaki	Corvinus University, Hungary
Frank Danielsen	University of Agder, Norway
Lieselot Danneels	Ghent University, Belgium
Athanasios Deligiannis	International Hellenic University, Greece
Gabriele De Luca	Danube University Krems, Austria
Monica Denboer	Università di Bologna, Italy
Edna Dias Canedo	Universidade de Brasília, Brazil
Devin Diran	TNO, The Netherlands
Bettina Distel	Universität Münster, Germany
Ioanna Donti	International Hellenic University, Greece
David Duenas-Cid	Gdansk University of Technology, Poland
Noella Edelmann	Danube University Krems, Austria
Gregor Eibl	Danube University Krems, Austria
Tove Engvall	University of Agder, Norway
Montathar Faraon	Kristianstad University, Sweden
Shahid Farooq	Government of the Punjab, Pakistan
Asbjørn Følstad	SINTEF, Norway
Marcelo Fornazin	FGV EBAPE, Brazil
Mary Francoli	Carleton University, Canada
Jonas Gamalielsson	University of Skovde, Sweden
Luz Maria Garcia	Universidad de la Sierra Sur, Mexico
Francisco García Morán	European Commission, Luxembourg
Mila Gasco-Hernandez	University at Albany, State University of New York, USA
Alexandros Gerontas	University of Macedonia, Greece

J. Ramon Gil-Garcia	University at Albany, State University of New York, USA
Dimitris Gouscos	University of Athens, Greece
Malin Granath	Linköping University, Sweden
Stefanos Gritzalis	University of Piraeus, Greece
Divya-Kirti Gupta	Indus Business Academy, India
Mariana Gustafsson	Linköping University, Sweden
Sebastian Halsbenning	Universität Münster, Germany
Marcus Heidlund	Mid Sweden University, Sweden
Moreen Heine	Universität zu Lübeck, Germany
Marissa Hoekstra	TNO, The Netherlands
Wout Hofman	TNO, The Netherlands
Sara Hofmann	University of Agder, Norway
Tomasz Janowski	Gdańsk University of Technology, Poland
Marijn Janssen	Delft University of Technology, The Netherlands
Marius Rohde Johannessen	University of South-Eastern Norway, Norway
Björn Johansson	Linköping University, Sweden
Luiz Antonio Joia Luiz	Getulio Vargas Foundation, Brazil
Hong Joo Lee	The Catholic University of Korea, South-Korea
Gustaf Juell-Skielse	Stockholm University, Sweden
Yury Kabanov	National Research University Higher School of Economics, Russia
Natalia Kadenko	Delft University of Technology, The Netherlands
Muneo Kaigo	University of Tsukuba, Japan
Evangelos Kalampokis	University of Macedonia, Greece
Nikos Karacapilidis	University of Patras, Greece
Evika Karamagioli	University of Athens, Greece
Areti Karamanou	University of Macedonia, Greece
Naci Karkin	Pamukkale University, Turkey
Jongwoo Kim	Hanyang University, South-Korea
Fabian Kirstein	Fraunhofer FOKUS, Germany
Jens Klessmann	Fraunhofer FOKUS, Germany
Bram Klievink	Leiden University, The Netherlands
Ralf Klischewski	German University in Cairo, Egypt
Michael Koddebusch	European Research Center for Information Systems, Germany
Robert Krimmer	University of Tartu, Estonia
Peter Kuhn	Fortiss, Germany
Zoi Lachana	University of the Aegean, Greece
Mariana Lameiras	United Nations University, Portugal
Thomas Lampoltshamme	Danube University Krems, Austria
Habin Lee	Brunel University London, UK

Azi Lev-On	Ariel University, Israel
Matthias Lichtenthaler	Bundesrechenzentrum, Germany
Johan Linåker	RISE Research Institutes of Sweden, Sweden
Katarina Lindblad-Gidlund	Mid Sweden University, Sweden
Ida Lindgren	Linköping University, Sweden
Nuno Lopes	DTx - Digital Transformation CoLAB, Portugal
Euripidis Loukis	University of the Aegean, Greece
Rui Lourenço	University of Coimbra, Portugal
Michalis Loutsaris	University of the Aegean, Greece
Tangi Luca	Joint Research Centre, European Commission, Spain
Luis F. Luna-Reyes	University at Albany, SUNY, USA
Bjorn Lundell	University of Skövde, Sweden
Ahmad Luthfi	Delft University of Technology, The Netherlands
Johan Magnusson	University of Gothenburg, Sweden
Heidi Maurer	Danube University Krems, Austria
Keegan McBride	Hertie School Centre for Digital Governance, Germany
John McNutt	University of Delaware, USA
Rony Medaglia	Copenhagen Business School, Denmark
Ulf Melin	Linköping University, Sweden
Sehl Mellouli	Laval University, Canada
Ana Melro	University of Aveiro, Portugal
Tobias Mettler	University of Lausanne, Switzerland
Morten Meyerhoff Nielsen	United Nations University, Portugal
Yuri Misnikov	University of Leeds, UK
Francesco Mureddu	The Lisbon Council, Belgium
Marco Niemann	European Research Center for Information Systems, Germany
Anastasija Nikiforova	University of Tartu, Estonia
Anna-Sophie Novak	Danube University Krems, Austria
Galia Novakova-Nedeltcheva	Politecnico di Milano, Italy
Hannu Nurmi	University of Turku, Finland
Ann O'Brien	NUI Galway, Ireland
Svein Ølnes	Vestlandsforsking, Norway
Monica Palmirani	University of Bologna, Italy
Panos Panagiotopoulos	Queen Mary University of London, UK
Peter Parycek	Danube University Krems, Austria
Samuli Pekkola	Tampere University, Finland
Rob Peters	HU University of Applied Sciences Utrecht, The Netherlands
Luiz Pereira Pinheiro Junior	Universidade Positivo, Brazil

Vigan Raca	Ss. Cyril and Methodius University of Skopje, North Macedonia
Michael Räckers	European Reserach Center for Information Systems, Germany
Luis Felipe Ramos	University of Minho, Portugal
Barbara Re	University of Camerino, Italy
Nicolau Reinhard	University of São Paulo, Brazil
Britta Ricker	Utrecht University, The Netherlands
Nina Rizun	Gdansk University of Technology, Poland
Manuel Pedro Rodríguez Bolívar	University of Granada, Spain
Alexander Ronzhyn	University of Koblenz-Landau, Germany
Boriana Rukanova	Delft University of Technology, The Netherlands
Per Runeson	Lund University, Sweden
Øystein Sæbø	University of Agder, Norway
Amir Sahi	Bern University of Applied Sciences, Switzerland
Rodrigo Sandoval-Almazan	Universidad Autonoma del Estado de Mexico, Mexico
Verena Schmid	Donau-Universität Krems, Austria
Hendrik Scholta	University of Muenster and ERCIS, Germany
Johanna Sefyrin	Linköping University, Sweden
Uwe Serdült	Ritsumeikan University, Japan
Kerley Silva	University of Porto, Portugal
Anthony Simonofski	Université de Namur, Belgium
Søren Skaarup	IT University of Copenhagen, Denmark
Leif Sundberg	Mid Sweden University, Sweden
Iryna Susha	Utrecht University, The Netherlands
Proscovia Svärd	Mid Sweden University, Sweden
Efthimios Tambouris	University of Macedonia, Greece
Lörinc Thurnay	Danube University Krems, Austria
Jean-Philippe Trabichet	HEG Genève, Switzerland
Andrea Trentini	University of Milan, Italy
Jolien Ubacht	Delft University of Technology, The Netherlands
Sélinde van Engelenburg	Delft University of Technology, The Netherlands
Colin van Noordt	Tallinn University of Technology, Estonia
Marco Velicogna	IRSIG-CNR, Italy
Gabriela Viale Pereira	Danube University Krems, Austria
Shefali Virkar	Danube University Krems, Austria
Gianluigi Viscusi	Imperial College LOndon, UK
Flurina Wäspi	Berner Fachhochschule Wirtschaft, Switzerland
Frederika Welle Donker	Delft University of Technology, The Netherlands
Guilherme Wiedenhöft	Federal University of Rio Grande, Brazil
Elin Wihlborg	Linköping University, Sweden

Peter Winstanley	Semantechs Consulting, UK
Stijn Wouters	Katholieke Universiteit Leuven, Belgium
Anja Wüst	Berner Fachhochschule, Switzerland
Maija Ylinen	Tampere University of Technology, Finland
Chien-Chih Yu	National Chengchi University, Taiwan
Thomas Zefferer	A-SIT Plus GmbH, Austria
Dimitris Zeginis	University of Macedonia, Greece
Qinfeng Zhu	University of Groningen, The Netherlands
Sheila Zimic	Mid Sweden University, Sweden
Anneke Zuiderwijk	Delft University of Technology, The Netherlands

Additional Reviewers

Lucana Estevez	San Pablo CEU University Foundation, Spain
Stanislav Mahula	KU Leuven, Belgium
Simon Hunt	University of Manchester, UK

Contents

Legal Informatics

Digital Society

E-Democracy and E-Participation

Voter Authentication in Remote Electronic Voting Governmental Experiences: Requirements and Practices

Adrià Rodríguez-Pérez[1,2](✉), Jordi Cucurull[1], and Jordi Puiggalí[1]

[1] Scytl Election Technologies, S.L.U., 08021 Barcelona, Spain
{adria.rodriguez,jordi.cucurull,jordi.puiggali}@scytl.com
[2] Universitat Rovira i Virgili, 43002 Tarragona, Spain

Abstract. How to ascertain that the person voting behind a computer or smartphone screen is actually who they claim to be remains one of the key challenges in remote electronic voting. Credentials to vote online can be shared, stolen, or traded. For this reason, it is generally argued that introducing remote electronic voting from uncontrolled environments for political public elections is only feasible as long as a robust infrastructure for the digital identification of voters (e.g., based on electronic identity documents, e-ID) is already in place. But is such a digital infrastructure for voter authentication a *sine qua non* condition for remote electronic voting? In this paper we assess how voters are authenticated in internet voting for political public elections in nine countries: Australia, Canada, Estonia, France, Mexico, Pakistan, Panama, Switzerland, and the United States of America (USA). To the best of our knowledge, this is the broadest comparative assessment of voter authentication methods in governmental remote electronic voting experiences. Our analysis reveals that the use of solely knowledge-based factors for voter authentication is the most common practice in these experiences. In most cases, a combination of several credentials is used (e.g., in Canada and Australia). Another alternative is to rely on a different combination of knowledge and ownership-based authentication methods that does not require neither e-IDs nor digital certificates (e.g., as in France and Mexico).

Keywords: Voter authentication · Remote electronic voting · Equal suffrage

1 Introduction

How to achieve both the "unique voter identification (only eligible voters can cast a vote and those only once) and anonymous vote casting (the voter must be anonymous when he [sic] casts a vote) at the same time"[1] (Volkamer 2009: 2) remains today one of the key

[1] It should be noted that it may not be necessary to guarantee the anonymity of encrypted votes as soon as they are cast. As a matter of fact, in certain cases it may be even necessary to maintain a link between the encrypted vote cast and the identity of the voter who has cast it until the decryption stage (e.g., when multiple voting is supported).

Published by Springer Nature Switzerland AG 2022
R. Krimmer et al. (Eds.): ePart 2022, LNCS 13392, pp. 3–18, 2022.
https://doi.org/10.1007/978-3-031-23213-8_1

challenges of remote electronic voting[2] from uncontrolled environments[3]. How could it be ascertained that the person voting behind the computer or smartphone screen is actually who they claim to be and not someone who has forced them to hand them their credentials? Or who has stolen them? Or who has bought them?

According to some commentators, introducing remote electronic voting from uncontrolled environments for political public elections is only feasible as long as a strong system for the digital identification for their citizens is already in place. For example, the former President of Estonia, Toomas Hendrik Ilves, recently noted that "[m]any countries realize that strong remote voter authentication is an immense practical problem that has to be dealt with before they can consider deploying any online voting system" (2016: xi) and then praised Estonia's electronic e-ID system.

In practice, however, a comparative assessment of the main countries using internet voting some years ago concluded that "Estonia is the only country with digital identification, while the two other countries [Norway and Switzerland] use(d) either existing physical IDs or wholly new unique identification methods valid only for each election" (Vinkel 2016: 48). Would Estonia remain the only country using eIDs for voter authentication in internet voting if we assessed governmental experiences today?

According to International IDEA's ICTs in Elections Database, at the time of writing there are 12 countries where Internet voting systems are used: Armenia, Australia (New South Wales[4]), Canada (Ontario[5]), Estonia, France, Mexico (Ciudad de México, CDMX[6]), New Zealand, Oman, Pakistan, Panama, Switzerland, and United Arab Emirates. We have assessed how voters are authenticated in these governmental experiences, with two exceptions: Oman and the United Arab Emirates, which can be considered illiberal or hybrid political contexts according to Romanov and Kabanov (2020) and thus

[2] Remote electronic voting is understood here as those systems "where votes are transferred via the Internet to a central counting server. Votes can be cast either from public computers or from voting kiosks in polling stations or—more commonly—from any Internet-connected computer accessible to a voter" (International IDEA 2011: 11). We will use the terms "remote electronic voting", "internet voting", and "online voting" indistinctively to refer to these systems.

[3] Whilst internet voting can be used from both controlled and uncontrolled environments, our focus here will be on the later (since it is in uncontrolled environments where it is more difficult to ascertain the identity and eligibility of a voter). In remote electronic voting from controlled environments (such as in polling stations, embassies and/or public libraries) a polling station officer could always verify, at least in principle, the identity of the voter.

[4] More recently, the Australian Capital Territory (ACT) has also introduced e-voting for voters overseas. This system was first used between 28 September and 17 October 2020.

[5] Online voting is also used, although to a lesser extent, in the Canadian province of Nova Scotia. However, and because of its size, we have decided to focus on Ontario: according to Cardillo, Akinyokun and Essex, "in the context of Ontario's 2018 municipal elections […] as many as one million voters cast a ballot online" (2020: 7).

[6] In addition to CDMX, 11 Mexican states used internet voting for the state elections of 6 June 2021, in partnership with Mexico's Election Management Body, the Instituto Nacional Electoral (INE): Baja California Sur, Chihuahua, Colima, Guerrero, Michoacán, Nayarit, Querétaro, San Luis Potosí y Zacatecas, Guerrero (*diputación migrante*) and Jalisco (*diputación representación proporcional*). More information can be found at: <https://www.dof.gob.mx/nota_detalle.php?codigo=5573949&fecha=01/10/2019> [retrieved: 18 March 2022].

do not necessarily comply with international standards for democratic elections[7]. In turn, no data has been found for Armenia nor for New Zealand. Additionally, we have also assessed voter authentication methods for the internet voting systems used in the United States of America (USA)[8], a country that is not included in International IDEA's database but where internet voting is used at different levels for political public elections and was piloted during the 2018 mid-term elections in the State of West Virginia.

To conduct this assessment, the following Sect. 2 first considers the requirements for voter authentication from an international perspective. Following, in Sect. 3 we provide a detailed assessment of voter authentication methods in each of these countries. To introduce these experiences, we rely on the classification from Volkamer (2009), who distinguished between voter authentication methods based on (a) something you know (i.e., knowledge), (b) something you have (i.e., ownership), and (c) something you are (i.e., biometrics), or (d) any combination of these. For each of these categories, we will provide detailed analysis for certain experiences that allow us to understand how and why these voter authentication methods are used and made more robust. The last section of the paper (4) provides an overview of the main methods used and offers a discussion of the advantages and limitations of each method. Lastly, the conclusion (5) highlights the contribution of this paper, acknowledges some constraints, and suggests some follow-up work.

This work is relevant for three main reasons. First, because it updates existing literature on voter authentication with the most recent practices in political public elections. Second, because it provides a comparative assessment of voter authentication methods, which offers guidance to researchers and to practitioners when it comes to understanding the bigger picture about voter authentication in remote electronic voting[9]. Third, our assessment is valuable because it offers an interdisciplinary approach to the topic of voter authentication.

2 Equal Suffrage and Voter Authentication: Legal Requirements

Volkamer rightly points out that "[e]very remote electronic voting system needs to implement voter identification and authentication techniques to ensure that only eligible voters

[7] For example, Oman has a score of 0.08 in the category of electoral process and pluralism in the Economist's Democracy Index (The Economist Intelligence Unit 2020), and the United Arab Emirates' score is 0.00. In contrast, Australia scores 10.00, Canada, Estonia, France, Panama and Switzerland score 9.58, Mexico scores 7.83, Armenia scores 7.50, and Pakistan scores 5.67. The same could be said of the Russian Federation, a country where internet voting has been used as well, but that is not included in International IDEA's database. Russia scores 2.17 in the category of electoral process and pluralism in the Economist's Democracy Index.

[8] The USA score 9.17 in the category of electoral process and pluralism in the Economist's Democracy Index (The Economist Intelligence Unit 2020).

[9] A recent study commissioned by the European Commission also offers a comparative assessment of voter authentication methods in governmental experiences with remote electronic voting (Lupiáñez-Villanueva and Devaux 2018). However, this study is not as comprehensive as our (several of our case studies are not considered, such as Mexico and New South Wales) and focuses more on providers than on actual experiences.

may cast a vote and those only once" (2009: 25). Certainly, the principle of equal suffrage requires that "[t]he principle of one person, one vote must apply, and within the framework of each State's electoral system, the vote of one elector should be equal to the vote of another" (Human Rights Committee 1996: 7). In Europe, for example, the European Commission for Democracy through Law – more commonly known as the Venice Commission – also identifies equal voting rights as one of the guidelines of democratic elections, ascertaining that "each voter has in principle one vote; where the electoral system provides voters with more than one vote, each voter has the same number of votes" (Venice Commission 2002: 6).

The Council of Europe's Recommendation Rec(2017)5 on standards for e-voting – which remains to date the only intergovernmental standard in the field[10] (see Driza Maurer 2017) – offers further guidance on how equal suffrage's requirements for voter authentication are to be understood when technology is introduced for the casting of the votes. For example, standard No. 7 of the Recommendation prescribes that "[u]nique identification of voters in a way that they can unmistakably be distinguished from other persons shall be ensured" (Council of Europe 2017a: 5). The Explanatory Memorandum to the Recommendation further details that "unique identification refers to validating the identity of a specific person by means of one or more features so that the person can unmistakably be distinguished from other persons" (Council of Europe 2017b: 7). The description for this standard also reads that "[a]uthentication can be identity-based and role-based [...] identity-based identification is advisable for voters registering or casting a vote" (Council of Europe 2017b: 7).

Additionally, standard No. 8 reads that "the e-voting system shall only grant a user access after authenticating her/him as a person with the right to vote" (Council of Europe 2017a: 5). Notwithstanding, the provisions in the Explanatory Memorandum for this standard clarify that "where anonymous voting tokens prove that a voter is eligible to vote, identification of the voter may not be required at this point as it has already taken place at an earlier stage, namely when the specific token is assigned to a specific voter" (Council of Europe 2017b: 8).

3 Voter Authentication in Practice: Governmental Experiences

In what follows, we classify the different experiences with remote electronic voting in political public elections based on the methods that they use for voter authentication. Throughout the paper, we will offer more specific description about some of these cases to illustrate the voter authentication methods they use.

Our analysis starts with Volkamer's three ways of identification and authentication as applied to remote electronic voting: something you know (knowledge), something you have (ownership), and something you are (biometrics) (Volkamer 2009: 25)[11]. A combination of any of these techniques is possible as well.

[10] Even if the geographic scope of the Recommendation is in principle limited to the countries of the Council of Europe, there are several examples of non-European countries resorting to them. See for instance Stein and Wenda (2014) and Driza Maurer (2014). More recently, Essex and Goodman (2020) have also assessed the Council of Europe's approach towards the regulation of e-voting as a model for the development of internet voting standards in Canada.

[11] Similar classifications are offered by Krimmer et al. (2007) and Abu-Shanab et al. (2013).

Our assessment identifies knowledge-based authentication methods as the most common form of voter authentication in governmental experiences, in most cases combining different secrets or credentials that can be delivered to voters using different channels (e.g., by e-mail, SMS, post or by phone). It is used in Australia (New South Wales), Canada (Ontario), Pakistan, Panama, and Switzerland. To the best of our knowledge, biometric identification has only been used in the USA, during the 2018 mid-term elections when the State of West Virginia conducted a pilot using a blockchain-based internet voting system. Ownership-based voter authentication systems are not used in any of the cases we analyse, but a combination of ownership- and knowledge-based authentication methods.

3.1 Knowledge-Based Voter Authentication Methods

Our analysis reveals that the use of knowledge-based factors for voter authentication is the most common practice in governmental remote electronic voting experiences. It is used in the Australian States of New South Wales (with iVote Number and password), the Canadian province of Ontario (PIN and date of birth), Pakistan (where voters registering to vote online are asked several questions considered secret[12]), Panama (C1V verification key[13]), and Switzerland (password and birthdate, access code, and access code with validation code, depending on the cantonal implementation). This analysis also shows that more than one secret is needed to authenticate and vote in most of these experiences. In some cases, these secrets are delivered to voters through different channels (e.g., by e-mail, SMS, post or by phone) to mitigate the risk of voter impersonation.

In what follows, we explain how knowledge-based voter authentication methods work in Switzerland, Ontario (Canada), and New South Wales (Australia):

Switzerland. Since Switzerland started piloting remote electronic voting already in 2003, three different systems have been used in elections at all levels (communal, cantonal, federal) and in referendums: the ones in Geneva, Neuchâtel, and Zurich (later known as the *consortium*). According to the Swiss Federal Chancellery, Switzerland has held more than 300 electoral events in which Swiss voters in up to 15 cantons have been able to vote online (2020: 3). Even if the use of internet voting was recently discontinued, the federal government has already started working on a new legal framework and it is expected that internet voting will be offered again soon. Since it is the responsibility of the cantons to implement internet voting, we have found several alternatives when it comes to voter authentication:

For example, in Geneva the principle that has guided the development of remote electronic voting is "simplicity" (Swiss Federal Council 2006: 5222). The remote electronic voting process had to be as similar as possible to the one in place to vote by post (Swiss

[12] For more information about this specific voting experience, the reading by Haq et al. (2019) is suggested.

[13] There is not a lot of information about how this system works. However, we have found this reference to the C1V verification key in a document by the ACE Project: <https://aceproject.org/ero-en/regions/americas/PA/panama-carpeta-informativa-elecciones-generales> [retrieved: 18 March 2022].

Federal Council 2002: 649; Swiss Federal Chancellery 2004: 8): voters received their credentials (a password) by post, shielded by a tamper-evident overlay. To reveal their password, they had to scratch the area of their voting card where the password had been hidden. To authenticate themselves, they had to introduce this password as well as their birthdate into the system. The team responsible for this pilot project thus considered that the voting card was at the core of their system (Swiss Federal Chancellery 2004: 35).

The project in Neuchâtel was framed within a wider digitisation process, the so-called *guichet virtuel* (Swiss Federal Council 2002: 650) or *sécurisé* (Swiss Federal Chancellery 2004: 8) *unique*. Each voter eligible to vote online (namely, those registered at the *guichet*) would receive a unique confidential voter card together with their voting materials ahead of a contest (Swiss Federal Council 2002: 651). This voter card contained unique references (a barcode) and a security hologram (Swiss Federal Chancellery 2004: 41) and a unique access code to access the voting platform within the *guichet* (Swiss Federal Council 2002: 651). To cast their ballot, a voter had to type the validation code provided in their voting card. Thus, a two-step authentication was used here: first to ascertain the identity of the voter, and then to confirm the eligibility of the vote cast. This could be considered equivalent to have an OTP embedded in the voter card (i.e., sent by postal instead of SMS).

In Zurich's system, as in the two other pilot projects, each eligible voter received their credentials ahead of an election or referendum. However, in this experience voters could identify themselves against the internet voting system by typing just an access code printed on their voter card.

Ontario, Canada. Municipalities in the Canadian province of Ontario have been using internet voting for its municipal elections since 2003. Elections in Canada are considered "highly digital" (Essex and Goodman 2020: 163), with "more than 4.5 million online 'voting opportunities' in these municipalities since 2003" (Goodman and Smith 2017: 167). While online voting experiences have in general been positive for stakeholders (Goodman and Smith 2017: 169), some have warned that in Canada there is a lack of "safeguards in place such as standards, guidelines, or bodies that provide certification to regulate electronic voting" (Essex and Goodman 2020: 163). According to Essex and Goodman, "Canada's multilevel governance structure has meant municipalities mostly deliver election on their own terms, resulting in a patchwork of online voting models and security requirements" (2020: 162). This patchwork extends to voter authentication methods.

For example, a recent report about the Ontario municipal elections of 2018 found "weak voter authentication" practices (Cardillo et al. 2020: 5). According to these authors, "[t]he primary credential needed to cast a ballot online consisted of a knowledge factor (a PIN and/or ID) transmitted to the voter in a voter information package via postal mail"[14] (Cardillo et al. 2020: 5). Due to this already-mentioned multilevel governance, these passwords ranged in length in each municipality from 9 digits (with Simply Voting, used in 28 municipalities) to 16 digits (with Intelivote and Scytl, used in 100 municipalities). These authors have also observed that the channel used to deliver the

[14] These authors identify one exception in the city of Cambridge, where PINs were sent by email (Cardillo et al. 2020: 5).

credentials could be compromised. According to their findings, "some voters observed that the PINs were legible through the envelope when held up to bright light" (Cardillo et al. 2020: 5).

Possibly in an effort to mitigate this risk, "[i]n almost all cases a second knowledge factor (date of birth) was required" to authenticate a voter (Cardillo et al. 2020: 5). While adding a second knowledge factor could seem as increasing the robustness of voter authentication[15], Cardillo et al. (2020: 5) argue that "[d]ates of birth, however, make a poor login credential [...] knowledge of a PIN or date of birth does not establish a voter's identity. It merely establishes to the voting server that some entity on the other end of the connection knows a secret. Secrets, of course, can be transferred or intercepted". Therefore, they conclude, "authentication is still considered single-factor (as opposed to multi-factor) authentication since both credentials are knowledge factors" (Cardillo et al. 2020: 5).

New South Wales, Australia. In Australia, certain groups of voters in the state of "New South Wales, the most populous of Australia's six states, are able to cast their votes via the internet or telephone using the iVote system" (Goodman and Smith 2017: 171). According to these authors, "since 2011, [New South Wales'] voters have cast nearly 339,000 votes across nine elections" (Goodman and Smith 2017: 171). To this number, one should add the online votes cast during the last 2019 State elections.

To vote with iVote, New South Wales' internet voting system, a voter has to follow an online registration process first in which they prove their identity introducing several personal data (e.g., birth name, surname or family name, postal code, location, street name and, optionally, a passport, driver licence or Medicare details as a secondary confirmation of identity)[16]. Then, the voters have to introduce their contact information, which includes a mobile phone number, an e-mail and/or telephone number. After selecting to vote online, voters have to introduce a password that will be used to authenticate them to vote at a later stage. Passwords are scored, and at least a score of 4 out of 5 has to be obtained for the password to be accepted (according to its length, combination of letters, numbers, etc.).

Each password is then associated to an iVote Number that represents the voter. The iVote Number is a numeric value of 8 digits and is sent to each voter through the channel of their choice (SMS, email, post, or telephone call). Both values, the iVote Number and the password, are used to authenticate a voter in the iVote system before they can cast their vote.

3.2 Biometric-Based Voter Authentication Methods

To the best of our knowledge, biometric authentication has only been used as a voter authentication method in governmental elections in the USA and, more specifically, in

[15] The authors highlight that "[t]he use of single credential for voter authentication is inadvisable since access to the voter information package is sufficient to cast a ballot on another's behalf" (Cardillo et al. 2020: 5). However, we have already seen that a single credential for voter authentication is the method that has been used in some cantons in Switzerland for both internet voting as well as postal voting.

[16] More information about this method can be found online: <https://www.elections.nsw.gov.au/Voters/Other-voting-options/iVote-online-and-telephone-voting> [retrieved: 18 March 2022].

West Virginia[17]. Although the use of this system has been limited (during the 2018 mid-term elections 114 voters from 31 countries used the system[18]), this approach to voter authentication is quite novel and deserves some specific attention. In what follows, we explain this experience with more detail:

West Virginia, the USA. As already mentioned, the States of West Virginia piloted the use of an internet voting application during the 2018 mid-term elections. It was the first time that voters in the USA could cast their ballot on a mobile phone (Specter et al. 2020: 1535). The possibility to vote online was offered to deployed members of the military and overseas citizens[19]. The mobile application used, provided by the company Voatz Inc., relied on blockchain technology and biometric voter authentication. More specifically, voter authentication worked in three steps[20]: (1) the voter scans their driver's license, ID or passport; (2) they take a live snapshot of their face (what Voatz calls a "video selfie"); and (3) touches the fingerprint reader or uses the facial recognition feature on their smartphone, which links the device to the voter (alternatively, this last step could be substituted by just introducing a 8 digit PIN).

According to Voatz, "[t]he app first does a liveness check on the 'selfie', then compares the voter's "selfie" to the photo on their passport or driver's license, and finally, compares the ID data to the state's voter registration database to confirm that the voter is eligible to vote". One of the advantages of such a biometric authentication method is that there is no need to store a central biometric template database of eligible voters beforehand, since the comparison is done between the picture in the voter's physical ID and the live snapshot. Additionally, it is possible to delete all the data once the voter has been authenticated.

3.3 Combinations: Ownership- and Knowledge-Based Voter Authentication Methods

As we have mentioned, a combination of the three ways of identification and authentication is also possible. In the case of internet voting, we have found three experiences in which voter authentication is done based on a combination of ownership - and knowledge-based voter authentication methods: Estonia, France, and Mexico. In these three cases, voter authentication is achieved by means of knowledge and ownership. In what follows, we explain these three voter authentication methods:

[17] Additionally, there are several proposals from academia and from industry for biometric voter authentication in remote e-voting. See for example Hof (2004) and Morales-Rocha et al. (2008). However, we are not aware that any of these proposals has been implemented in actual governmental experiences.

[18] Due to security concerns and vulnerabilities found by a group of researchers (see Specter et al. 2020), the system was not used during the 2020 presidential and legislative elections.

[19] More information about the results of the pilot can be found online: <https://sos.wv.gov/news/Pages/11-16-2018-A.aspx> [retrieved: 18 March 2022].

[20] More information about this method can be found online: <https://voatz.com/wp-content/uploads/2020/07/voatz-security-whitepaper.pdf> [last accessed: 18 March 2022].

France. Internet voting in France dates back to 2003, with the passing of the first law allowing the use of internet voting for the elections to the High Council of French Citizens Abroad (Anziani and Lefèvre 2014: 38). Nowadays, remote electronic voting is foreseen as an additional voting channel for French voters abroad: they can cast a remote electronic vote for the elections to the National Assembly and for the election of the Consular Advisers and Delegates. In 2012, voters had the possibility to vote online for 11 seats at the National Assembly (Anziani and Lefèvre 2014: 37). However, in 2017 this possibility was halted due to concerns of foreign cyber threats as well as over certain technical issues (Deromedi and Détraigne 2018: 35–36). On their side, Consular Advisers and Delegates are based at each embassy with a consular district and at each consular post. They are elected for a six-year mandate during the month of May, their first elections taking place in 2014 (Anziani and Lefèvre 2014: 37).

In France, the Electoral Code describes quite in detail the voter authentication method for internet voting. In this sense, Article R176-3-7 of the Code reads that the voter's identity when voting online has to be satisfied by means of an *identifiant* (username) and a *mot de pass* (password). These two credentials, which according to the aforementioned article cannot be linked to the civil status of the voter, are created randomly and delivered to them using two different channels[21]. The length of the credentials is 12 characters each with an alphabet of 58 symbols. Following the provisions of the Electoral Code, these credentials are randomly generated by a module of the voting system devoted to this purpose. Two ceremonies are held to create these credentials, one to start the key generation and another one to finish it[22], within a secure room. For the secure delivery of these credentials, the *identifiant* is protected with a One Time Secret (OTS) and delivered to voters in a personalised e-mail. The *mot de pass* is delivered by SMS, in clear text.

Up until here, this authentication method would seem no different from the ones we have just explained. However, for the voter to be able to confirm their choices, a third secret is needed. In order to access the voting platform, the voters identify themselves introducing just the *identifiant* and the *mot de pass*. This authentication method is enough for them to select the voting options and cast their vote. However, the vote is confirmed only if they can provide a third credential. Upon casting their vote, voters receive a One Time Password (OTP) of six digits by e-mail. The voter has to open the OTP and introduce it into the voting website in order to confirm the casting of the vote. If the password is not introduced, the vote is not confirmed. Therefore, since and e-mail account is also needed to confirm the vote, an ownership-based method is introduced.

CDMX, Mexico. Internet voting has been used in CDMX since 2010. This option has been offered to voters abroad for the election of the Head of Government of the City of México (in 2012, when both online voting and postal voting were available) and for several participatory processes (Chorny 2020: 60–61). Internet voting has been used recently in two participatory processes, from 8 to 15 March 2020: for the election of

[21] The article also reads that the credentials are to be delivered to voters at the opening of the voting phase at the latest and at least one of the two credentials is needed to recover the other one in case of loss.

[22] There are two ceremonies because the process takes many hours (for the consular elections of 2021, about 1.3 million credentials are needed), and it is not suitable to have the observers during all the time.

different citizen bodies (*Comisiones de Participación Comunitaria*), and for its participatory budgets for 2020 and 2021. Internet voting was used as an advanced voting channel (8 to 12 March) and in four polling stations on election day (15 March). In 2012, CDMX used the internet voting system of a private vendor (Scytl). After this project, the city's election administration decided to develop its own system.

With this new system, as in New South Wales, voters willing to vote online in CDMX have to register beforehand. Voters can register in person or online. They can register online by introducing their voting credentials (voter key and a code called *Código de Reconocimiento Óptico de Caracteres*, OCR). The introduction of these credentials can be done by taking a picture to the voter card or by typing them manually. After introducing these credentials, they have to register a mobile phone number and choose a channel for the delivery of an 8-digit password (to choose between e-mail or post). If they choose to have their password delivered by e-mail, face recognition techniques are used to ascertain the identity of the voter[23].

To vote, each voter has to introduce their voter key and the OCR by taking a picture of their voter card, as well as the password that was handed to them (when registering in person) or delivered by e-mail or post. If the three credentials are correct, the voter then receives an SMS to the phone number they registered with a 6-digit key. Only when they introduce this last key, they can access the voter portal and vote.

Estonia. Estonia remains, to date, the only country where all voters are offered the possibility to vote online, and at all levels: for municipal and parliamentary elections, as well as for elections to the European parliament and in referendums. With online voting being offered since 2005, the share of i-voters has increased steadily and in the last 2019 parliamentary elections about 44% of all votes were cast electronically (Heiberg et al. 2020: 82). The success of Estonia's internet voting experience cannot be divorced from the country's digital agenda, at whose core lies the e-ID. After passing the Identity Document Act in 1999 and the Digital Signature Act in 2000, the first ID cards were issued in January 2002. Between 2002 and 2014, "about 1.2 million of these credit-card size personal identification documents have been issued, allowing citizens to digitally identify themselves and sign documents or perform actions" (Vassil 2016: 16). Nowadays, Estonia's citizens and residents can use the mobile version of the electronic identity document (ID) to prove who they are (Mobile-ID), as well as the Digi-ID.

Therefore, to vote online in Estonia "Identity document (ID card), Mobile-ID, and digital identity document (Digi-ID)[24] can be used as tools for giving digital signature" (State Electoral Office of Estonia 2017: 6). According to the country's State Electoral Office, "most of the persons who have the right to vote possess an ID card that enables

[23] While the resort to these techniques could imply that biometrics are used for voter authentication, we have not considered face recognition as a method used for voter authentication *as such* (since the use of these techniques is limited to the registration phase and they are not used at the time of voting). In a similar vein, Barrat Esteve and Morales-Rocha have noted that the authentication process could be strengthened if biometrics were used at the time of voting instead (2020: 8).

[24] However, according to Heiberg, Krips, and Willemson, "[r]ight now, only ID-card and mID are used for i-voting" (2020: 83).

secure electronic identification and giving digital signature; many people also have an additional legally backed electronic ID document, like Digi-ID or Mobile-ID" (2017: 4). In this sense, and while Estonia's internet voting system "supports a variety of authentication methods that the Voter can choose from depending on the authentication credentials at his [sic] disposal" (State Electoral Office of Estonia 2017: 11) such as knowledge-based methods (e.g., username, password, PIN), "stronger identification security is ensured by a physical authentication token (e.g., chip card, SIM-card, etc.) combined with a knowledge-based PIN" (State Electoral Office of Estonia 2017: 11).

4 Discussion

All in all, it is clear that the main authentication method used is knowledge-factors (see Table 1). These methods may be used as the unique factor, or in combination with others (mainly ownership-based methods). Biometrics has only been used in one of the experiences that have been analysed.

Table 1. Voter authentication methods in each governmental experience

	Australia, New South Wales	Canada, Ontario	Estonia	France	Mexico, CDMX	Pakistan	Panama	Switzerland	The USA, West Virginia
Knowledge-based	• (2)	• (2)	• (2)	• (2)	• (3)	•	•	•	
Ownership-based			• A	• B	• C				
Biometric-based									•

( ) number of secrets that compose the voter credential.
A Identity document (ID card), Mobile–ID, and digital identity document (Digi–ID).
B OTP delivered by e–mail.
C OTP delivered by SMS.

As we have seen, there are important variations on how voter authentication is based on knowledge factors in these experiences. In some cases, just one knowledge factor has been used for voter authentication (as in Zurich, Switzerland). However, in the majority of cases two different credentials are required. Our analysis shows that adding a second credential does not necessarily prevent impersonation, since it is always possible to share this second secret as well (and sometimes the two credentials are delivered together, which makes it very easy for them to be shared or traded).

Notwithstanding, our assessment reveals that there are different ways in which secrets can be delivered to a voter. For instance, in New South Wales voters can choose their own password (as long as it meets some security requirements) and then have their username (i.e., iVote Number) delivered to them through the channel of their choice (e.g., SMS, email, post, or telephone call). In Ontario some information about the voter is used as this second secret (i.e., their birthdate).

In all circumstances, it seems that while highly usable and with low costs (although these can increase depending on how the credentials are delivered to voters), knowledge-based authentication is not the most secure choice. As Volkamer already noted, with knowledge-based voter authentication "vote buying cannot be excluded, because voters could easily send electronically their login data to a potential buyer" (2009: 25). At the end of the day, this risk will have to be evaluated on a case-by-case basis (e.g., in Switzerland, where this risk may be higher if compared to the other experiences, the authorities seem to have accepted it since it is similar to the risk of voter impersonation in postal voting, which relies on the same authentication method).

Overall, biometric-based voter authentication has the key advantage of assuring that the voters are who they claim to be (since biometrics cannot be shared or traded). However, it is also important to notice that there may be errors in the biometric technologies used which could prevent an authentic voter from authenticating themselves or allowing a non-eligible voter to access the voting platform. On the other hand, concerns about cost as expressed by Volkamer when she noted that "[l]argescale biometric infrastructures do in general not yet exist" (2009: 26) are to some extent mitigated if biometric voter authentication is implemented as it was done in West Virginia (since there is no need for the Electoral Management Body to keep biometric templates for all the electorate). Notwithstanding, this method restricts the devices that can be used to vote and increases dependence on third parties. For instance, Specter, Koppel and Weitzner noted that in Voatz's internet voting system, "[t]he user verifies their identity, using Voatz's integration with a third-party service called Jumio" (2020: 1538). Thus, not all costs can be overcome.

Lastly, combination of methods deserves a specific mention. In Mexico and France, an OTP is delivered to the voter at the time of voting, provided that they have introduced the proper credentials. In the French case, ownership is based on email addresses. While this method may have some advantages in terms of robustness, it is also important to highlight that the choice of e-mail for the delivery of the OTP may not compensate for the additional complexity added to the casting of the vote. In this sense, it does not seem unfeasibly to share the email account where the OTP is going to be received with someone else, since e-mail accounts can be opened from any device with a browser and internet connection, and the same e-mail account can have several concurrent sessions opened in different devices. The Mexican alternative, where an OTP is received by SMS, seems more robust[25].

Lastly, voter authentication based on tokens is not ideal either (in spite of Estonia's combination seems most robust and secure if compared to the other ones). It offers more security when it comes to ascertaining the identity of the voter, but physical tokens and passwords can be traded as well, and vote-buying cannot be completely excluded. More

[25] At the same time, delivering the OTP using another channel (such as SMS), which would require sharing the mobile phone or at least the SIM-card, does not seem suitable either, since internet voting is offered to French voters abroad and the delivery of the OTP would be subject to certain constraints that could render voting online almost impossible (i.e., if the SMS is not received before the voting session expires, the vote cannot be confirmed and thus remains not cast). This is not unimportant in view that up to 50% of voters involved in a User Acceptance Test (UAT) ahead of the 2017 French legislative elections had connection problems, mostly due to issues with their e-mail and the delivery of the SMS (Deromedi and Détraigne, 2018: 40). Issues with SMS were especially concerning in countries like China (Deromedi and Détraigne, 2018: 31).

recently, it has been highlighted that an attacker could submit a vote using a compromised e-ID environment without the voter noticing "where ID-card is left attached to the working terminal for extended periods of time, e.g. as a login token" (Heiberg et al. 2020: 84)[26]. Furthermore, this method is not as usable (at least when physical cards and associated card readers are required) and the costs are considerable. Estonia seems to have overcome this usability and infrastructure costs in the long term, as the e-ID is extensively used in the country and the internet voting system also supports Mobile-ID (and, according to the State Electoral Office of Estonia, Digi-ID). Notwithstanding, this method does not seem optimal for voter authentication unless the physical or digital tokens are extensively used[27]. The Mexican alternative, which relies on a mobile phone to deliver a third secret during the authentication phase (and which thus requires having at least the SIM-card at the time of voting) seems more feasible. However, as we have seen, it has its own shortcomings as well.

5 Conclusion

In this paper, we have aimed at assessing different methods for voter authentication in governmental remote electronic voting experiences. To do so, we have analysed international standards for voter authentication and their practical implementation in different countries. Our analysis sits somewhere in between a legal and a technological study. However, and because of the constrains of the format in which it is presented, it does not offer a deep assessment neither of the technological nor the legal aspects of these experiences. This has not been our goal either. Instead, we aimed at (1) clarifying which methods exist and are used to ensure that only eligible voters vote in governmental experiences when remote electronic voting is introduced, and (2) to understand which mechanisms can be put in place to make these methods more robust. We are aware that national differences in electoral policy impose different requirements for voter authentication, and thus we have not been willing to evaluate all these methods against abstract standards and requirements.

Notwithstanding, our contribution is valuable for several reasons. First, because it updates existing literature on voter authentication with the most recent practices. Second, because it provides a comparative assessment which can offer guidance to researchers and to practitioners when it comes to understanding the bigger picture about voter authentication in remote electronic voting. This comparative assessment has also allowed us to challenge some conventional wisdom about the necessary pre-conditions for the introduction of internet voting: that an infrastructure for the digital identification of voters is a *sine qua non* condition and the best way to achieve voter authentication. As we have seen, only one country in our analysis uses digital identification cards (Estonia). By contrast, most of countries we have analysed have opted: (1) for analogous methods to those already existing for paper-based voting channels, such as postal voting (i.e.,

[26] This vulnerability, described as the Ghost Click Attack by Springall et al. (2014), was first identified in 2013.

[27] While not very common for governmental experiences, it is not unlikely that such digital identification infrastructures could exist for private settings, such as in universities, political parties or professional associations, to name just a few examples.

Switzerland), (2) for more than one knowledge factor (e.g., Ontario in Canada), sometimes delivered to voters using different channels (e.g., New South Wales in Australia), or (3) for a combination of knowledge- and ownership-based authentication methods, without requiring e-IDs or digital certificates (e.g., France and Mexico).

Lastly, our assessment is valuable because it offers an interdisciplinary approach to the topic of voter authentication. This is important because, and while requirements for voter authentication tend to be formulated in similar ways in international standards and national regulations (with exceptions, such as France), our research shows that the scope of alternatives for Election Management Bodies to choose among is quite broad. This is relevant for two main reasons. On the one hand, countries where remote electronic voting is regulated can introduce changes to the mechanisms used for voter authentication without having to start lengthy and complex processes to amendments their legislation. On the other hand, countries willing to introduce internet voting can regulate the requirements for voter authentication in a way that is broad enough to accommodate those mechanisms that are more suitable at a later stage (i.e., by means of executive orders or administrative acts).

Acknowledgements. This work received support from the mGov4EU project, which has received funding from the European Union's Horizon 2020 research and innovation programme under grant agreement No. 959072.

References

Abu-Shanab, E., Khasawneh, R., Alsmadi, I.: Authentication mechanisms for E-Voting. In: Saeed, S., Reddick, C.G. (eds.) Human-Centered System Design for Electronic Governance, pp. 71–86 (2013)

Anziani, A., Lefèvre, A.: Vote électronique: preserver la confiance des électeurs. Rapport d'information fait au nom de la commission des lois (2014)

Barrat Esteve, J., Morales Rocha, V.M.: Informe de Voto Electrónico (2020)

Cardillo, A., Akinyokun, N., Essex, A.: Online voting in Ontario municipal elections: a conflict of legal principles and technology? Whisper Lab Research Report, Western University (2020)

Chorny, V.: El voto por internet en México: La libertad y la secrecía del voto condicionadas. R3D. Red en Defensa de los Derechos Digitales, Ciudad de México (2020)

Council of Europe: Recommendation CM/Rec(2017)5 of the Committee of Ministers to member States on standards for e-voting (2017a)

Council of Europe: Explanatory Memorandum to Recommendation CM/Rec(2017)5 of the Committee of Ministers to member States on standards for e-voting (2017b)

Deromedi, J., Détraigne, Y.: Réconcilier le vote et les nouvelles technologies. Rapport d'information fait au nom de la commission des lois (2018)

Driza Maurer, A.: Ten years Council of Europe Rec(2004)11. Lessons learned and outlook. In: Krimmer, R., Volkamer, M. (eds.) Proceedings of Electronic Voting 2014 (EVOTE 2014), pp. 111–117. TUT Press, Tallinn (2014)

Driza Maurer, A.: Updated European standards for E-voting. In: Krimmer, R., Volkamer, M., Braun Binder, N., Kersting, N., Pereira, O., Schürmann, C. (eds.) E-Vote-ID 2017. LNCS, vol. 10615, pp. 146–162. Springer, Cham (2017). https://doi.org/10.1007/978-3-319-68687-5_9

Essex, A., Goodman, N.: Protecting electoral integrity in the digital age: developing E-voting regulations in Canada. Election Law J. **19**(2), 162–179 (2020)

French Electoral Code

Goodman, N., Smith, R.: Internet voting in sub-national elections: policy learning in Canada and Australia. In: Krimmer, R., et al. (eds.) E-Vote-ID 2016. LNCS, vol. 10141, pp. 164–177. Springer, Cham (2017). https://doi.org/10.1007/978-3-319-52240-1_10

Haq, H.B., McDermott, R., Ali, S.T.: Pakistan's internet voting experiment. In: Krimmer, R., et al. (eds.) Fourth International Joint Conference on Electronic Voting E-Vote-ID 2019, Lochau/Bregenz, Austria, 1–4 October 2019. Proceedings. TalTech Press, Tallin (2019)

Heiberg, S., Krips, K., Willemson, J.: Planning the next steps for Estonian Internet voting. In: Krimmer, R., et al. (eds.) Fifth International Joint Conference on Electronic Voting E-Vote-ID 2020, 6–9 October 2020. Proceedings. TalTech Press, Tallin (2020)

Hof, S.: E-voting and biometric systems. In: Electronic Voting in Europe - Technology, Law, Politics and Society, Workshop of the ESF TED Programme Together with GI and OCG, Schloß Hofen/Bregenz, Lake of Constance, Austria, 7–9 July 2004. Proceedings, pp. 63–72 (2004)

Human Rights Committee: United Nations: General Comment No. 25 (1996)

Ilves, T.H.: Foreword. In: Solvak, M., Vassil, K. (eds.) E-Voting in Estonia: Technological Diffusion and Other Developments Over Ten Years. Johan Skytte Institute of Political Studies, in cooperation with Estonian National Electoral Committee, Tartu and Tallin (2016)

International IDEA: Introducing Electronic Voting: Essential Considerations (2011)

Krimmer, R., Triessnig, S., Volkamer, M.: The development of remote E-voting around the world: a review of roads and directions. In: Alkassar, A., Volkamer, M. (eds.) Vote-ID 2007. LNCS, vol. 4896, pp. 1–15. Springer, Heidelberg (2007). https://doi.org/10.1007/978-3-540-77493-8_1

Lupiáñez-Villanueva, F., Devaux, A. (eds.): Study on the Benefits and Drawbacks of Remote Voting. European Commission, Brussels (2018)

Morales-Rocha, V., Puiggalí, J., Soriano, M.: Secure remote voter registration. In: Proceedings of 3rd International Symposium on Electronic Voting (EVOTE 2008), Bregenz, Austria, 7–9 August 2008, pp. 95–108 (2008)

Romanov, B., Kabanov, Y.: The oxymoron of the internet voting in illiberal and hybrid political contexts. In: Krimmer, R., et al. (eds.) E-Vote-ID 2020. LNCS, vol. 12455, pp. 183–195. Springer, Cham (2020). https://doi.org/10.1007/978-3-030-60347-2_12

Specter, M.A., Koppel, J., Weitzner, D.: The ballot is busted before the blockchain: a security analysis of Voatz, the first internet voting application used in U.S. federal elections. In: 29th USENIX Security Symposium (USENIX Security 2020), pp. 1535–1553 (2020)

Springall, D., et al.: Security analysis of the Estonian internet voting system. In: Proceedings of the 21st ACM Conference on Computer and Communications Security (CCS 2014), November 2014

State Electoral Office of Estonia: General Framework of Electronic Voting and Implementation thereof at National Elections in Estonia (2017)

Stein, R., Wenda, G.: The Council of Europe and e-voting: history and impact of Rec(2004)11. In: Krimmer, R., Volkamer, M. (eds.) Proceedings of Electronic Voting 2014 (EVOTE 2014), pp. 105–110. TUT Press, Tallinn (2014)

Swiss Federal Chancellery: Le vote électronique dans sa phase pilote - Rapport intermédiaire (2004)

Swiss Federal Chancellery: Restructuration et reprise des essais. Rapport final du Comité de pilotage Vote électronique (CoPil VE) (2020)

Swiss Federal Council: Rapport sur le vote électronique du 9 janvier 2002: Chances, risques et faisabilité (2002)

Swiss Federal Council: Rapport sur les projets pilotes en matière de vote électronique (2006)

The Economist Intelligence Unit: Democracy Index 2020: In sickness and in health? (2020)

Vassil, K.: The Estonian e-government ecosystem. In: Solvak, M., Vassil, K. (eds.) E-Voting in Estonia: Technological Diffusion and Other Developments Over Ten Years. Johan Skytte Institute of Political Studies, in cooperation with Estonian National Electoral Committee, Tartu and Tallin (2016)

Venice Commission: Code of Good Practice in Electoral Matters: Guidelines and Explanatory Report (2002)

Vinkel, P.: Historical development and legal aspects. In: Solvak, M., Vassil, K. (eds.) E-voting in Estonia: Technological Diffusion and Other Developments Over Ten Years. Johan Skytte Institute of Political Studies, in cooperation with Estonian National Electoral Committee, Tartu and Tallin (2016)

Volkamer, M.: Evaluation of Electronic Voting: Requirements and Evaluation Procedures to Support Responsible Election Authorities. Springer, Heidelberg (2009). https://doi.org/10.1007/978-3-642-01662-2

Using Open Government Data to Facilitate the Design of Voting Advice Applications

Daniil Buryakov[1(✉)], Mate Kovacs[1], Victor Kryssanov[1], and Uwe Serdült[1,2]

[1] College of Information Science and Engineering, Ritsumeikan University, 525-8577 Kusatsu, Japan
gr0527ri@ed.ritsumei.ac.jp, kovacsm@fc.ritsumei.ac.jp, kvvictor@is.ritsumei.ac.jp

[2] Center for Democracy Studies Aarau (ZDA), University of Zurich, 8006 Zurich, Switzerland
uwe.serdult@zda.uzh.ch

Abstract. In the process of statement selection for online voting advice applications (VAAs) a considerable amount of time is spent for analyzing the domestic and foreign policies of a given country. However, harnessing large amounts of available open data, which would be useful in this design process, manually is impractical. In order to facilitate such time-consuming and labor-intensive work, the authors propose a system to assist VAA designers formulating policy statements. Using advanced language modeling and text summarization techniques and based on open government data related to politics during the legislature preceding an election, the system produces suggestions applicable for revising or creating new VAA policy statements. Experiments conducted on VAA and e-petition data from Taiwan show that the proposed system can generate meaningful suggestions for VAA designers and could therefore help reducing the cost of the VAA design process.

Keywords: Voting advice applications · Topic modeling · Text summarization · e-petition data · Open data

1 Introduction

Over the last two decades, in many democracies around the globe, it has become customary for elections to be accompanied by civic-tech tools providing potential voters with online information about political parties or individual candidates. The basic functionality can be described in analogy to a recommender system, in which the user's political preferences are matched with the politicians or political parties on offer. The matching is based on a set of policy statements and respective answers by the users on one side and the candidates or political parties on the other side. Historically, such voting advice applications (VAAs)

Published by Springer Nature Switzerland AG 2022
R. Krimmer et al. (Eds.): ePart 2022, LNCS 13392, pp. 19–34, 2022.
https://doi.org/10.1007/978-3-031-23213-8_2

started off on paper in the Netherlands but for obvious reasons very soon spread across the globe as online versions. For political scientists, computer scientists as well as information scientists, VAAs provide for an ideal playground to conduct interdisciplinary research. A relatively large body of scholarly literature on the subject already exists and is summarized in overviews [2,9,17]. The scholarly literature making use of the relatively easily accessible VAA-generated data is also steadily increasing. One of the appealing parts of VAAs are the rather large data sets they tend to generate, sometimes even across several countries, e.g. on the occasion of European Parliamentary elections [21,23]. Without going into all the pros and cons of VAAs as such it seems to be clear from the outset that to a large degree the quality of a VAA stands or falls with the quality of the policy statements being used. These policy statements represent the core of the application. They must on one hand capture the current political debate in a comprehensive manner and on the other hand span across the political spectrum regarding electoral candidates or political parties. In addition, they must read well and be easy to understand for all potential users. Formulating them is not a trivial task. It requires expertise and in depth knowledge on a constituency's politics, electoral system as well as on political party system. Besides VAA statement formulation and selection, the annotation of the VAA statements for all running candidates or political parties is an additional costly job to perform before a VAA can be launched [10]. Annotators have to either contact political parties directly or need to have the expertise to make that annotation on their behalf, taking on their perspective. However, this paper only deals with the first of the VAA related tasks typically performed by humans. The goal of this study is, therefore, to increase the efficiency of VAA formulation and statement selection by means of information technology and engineering.

Typically, VAA designers use party manifestos and websites, voter and candidate questionnaires as well as surveys to develop appropriate VAA policy statements on which political parties are taking a stance. To stress it once more, this is a time-consuming and labor-intensive process that can take several weeks for a whole team of researchers. In addition to evaluating how specific responses affect results, the designers must assess the importance of different political issues and select which topics are the relevant ones for an upcoming election. While there is a vast amount of politics-related text available online that could be utilized to aid VAA designers in statement formulation, manual assessment of such large corpora of text data is unfeasible. Therefore, there is a need for computational tools to process and analyze large volumes of politics-related text data to extract meaningful information for the VAA development and improvement.

This study suggests a system to assist VAA designers in formulating and selecting policy statements. Using politics-related textual data, the system produces a set of VAA policy statements to choose from. Experiments with the system's developed prototype have been conducted using e-petition data from the Taiwanese Join Platform [15], combined with VAA policy statements from the 2020 Taiwanese presidential election. Technical methods implemented in the system rely on a language representation model called BERT [7] (Bidirectional Encoder Representations from Transformer), which is known for the ability to take into account the context of a word. The BERT topic modeling technique and an extractive summarizer are used to both select the topically most representative petitions and to summarize them. Semantic similarity between the summaries and VAA statements is calculated to find the most applicable petition summaries. VAA designers can use the result to improve already existing VAA applications. The proposed system does not entirely eliminate the need for human assessment. However, the obtained results demonstrate that valuable suggestions for revising or creating new VAA policy statements can be produced.

The rest of this paper is structured as follows. Related work is presented in Sect. 2, while Sect. 3 describes the proposed methodology in more detail. Section 4 introduces the data used and experiments made with the preliminary results obtained. Section 5 presents the main findings and a discussion. In Sect. 6, conclusions and outlines for future work can be found.

2 Related Work

Using computer science, big data or machine learning approaches for enhancing citizen participation [4] and rendering parts of VAAs in a more dynamic or complete manner becomes more common. If we count the paper version as the first in a series of VAA technology generations, there currently seems to be a transition to a third generation of VAAs. While the second generation of VAAs went online and provided for a blunt matching of users and political actors during the pre-electoral phase, a smarter, more dynamic generation of VAAs is appearing. [28], for example, are using Twitter data to update the political topics that are currently under debate and should therefore be represented in a VAA. In addition, in combination with sentiment analysis, Twitter data is used to apply a weight for identified topics to individual candidates. Also, the use of topic modeling approaches with e-petition data as textual input was recognized to create valid proxies of citizens' policy suggestions [13]. Natural language processing was mainly used to help categorize the vast amounts of textual data. However, using openly accessible e-petition content to update or create VAA statements has been not previously explored yet.

With the advent of machine learning frameworks for natural language processing, a variety of text analysis methods was created. Topic modeling with BERT to analyze politics-related issues was applied in several studies. Da Silva et al. [26] analyzed 1,313 political comments regarding the two legislative proposals publicly available from the Brazil's Chamber of Deputies online platform. The authors demonstrated the applicability of topic modeling tools for discovering latent topics in comments and showed how this could lead to increased citizen engagement regarding governmental policies. In another study, 12 million tweets from August 1st, 2020 to September 15th, 2020 mentioning "QAnon" in the context of the US elections were investigated by means of topic modeling [3]. From the topic descriptions with word frequencies, it was found that only a few topics were dominant among the tweets, with the overwhelming support of one candidate and this candidate's ideas.

Contextualized embeddings from BERT for similarity calculation and summarization are widely used in fake news detection as well as in the political domain. Gaglani et al. [8] explored automating claim reliability analysis for messenger applications. BERT embeddings were used to measure the similarity between claim messages regarding major events and news articles from trusted sources related to the same topic. In another study [11], the authors proposed a multimodal fake news detection model that combines visual, textual, and semantic information. Semantic information was derived by calculating the cosine similarity between embeddings of textual and visual information. While the textual information is the news title embedded by a pre-trained BERT model, the visual information is from the article image vectorized by the VGG-16 model. In the final step, these embeddings were concatenated and fused with the attention layer to predict whether online articles were fake. This approach achieved better results than other baseline models. Shirafuji et al. [25] created a pre-trained BERT model specifically for the Japanese political domain using 27,078 statements from the Tokyo Metropolitan Assembly Minutes dataset. The built model was utilized to summarize new political utterances. The authors concluded that summarization performance would be improved by additional training on a news corpus or Wikipedia.

3 Methodology

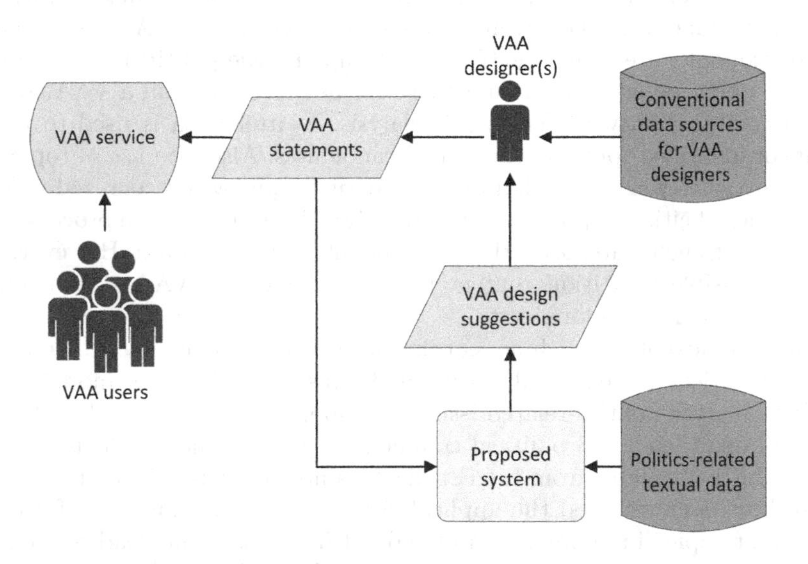

Fig. 1. The framework for the proposed system.

The proposed system has two inputs: VAA statements from the designers and textual data related to politics of the given country from the election campaign period (e.g., petition descriptions, social media data, blog posts, etc.). Figure 1 shows the framework for the system. Designers compile statements for the VAA service utilizing conventional data sources, such as party manifestos, candidate websites, surveys, etc. Based on the VAA design suggestions output by the proposed system, the designers potentially revise the original statements before deploying the VAA service on a web-based platform where it would be available to the users. Extracted from the politics-related data, the suggestions come in a textual format and have a complementary function to the already existing VAA policy statements. Thus, the proposed system neither aims to replace the original VAA statements and questions completely nor does it try to fully automate the job of VAA designers. The system produces design suggestions that would be either unpractical or unfeasible to produce manually.

3.1 Proposed System

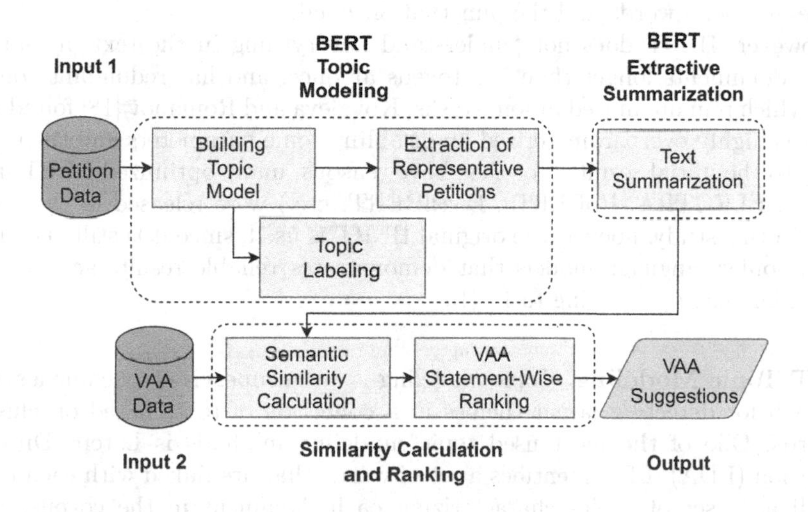

Fig. 2. The processing flow of the proposed system.

The detailed processing flow of the proposed system is given in Fig. 2. Since petition data is often openly available from governments, this study considers e-petition corpora as the politics-related textual data. Thus, in the figure, Input 1 and Input 2 are petitions and VAA statements, respectively (to produce meaningful results, both data have to be from the same country or region, ideally comprising the same legislature). Topic modeling is applied on the petition data to discover hidden topics and derive petitions that are topically the most representative of the data collection. After automatically summarizing the most representative petitions, the semantic similarity between the petition summaries and VAA statements are calculated to produce a ranked list of statement-wise suggestions ready to be further analyzed by domain experts and VAA designers.

The methods used to rely on the state-of-the-art language model Bidirectional Encoder Representations from Transformers (BERT) [7] that produces contextualized word embeddings. A word embedding is a numerical vector representing a word in a continuous vector space, where semantic similarity is calculated as the distance between word vectors. Unlike conventional word embedding models that produce static vectors like Word2Vec, BERT involves the bidirectional training of a transformer attention model, processing inputs in both directions. This helps better integrate the context into the vectors and makes the vectors contextualized. A pre-trained BERT model produces unique word vectors depending on the given input sentence, so the same words can have different vector representations based on their actual context. During the training, the model tries to predict the next sentence and a randomly hidden portion (10 to 15 percent) of tokens. Tokens are single units of meaning in a sentence. Tokens can be numbers, words and the punctuation itself.

However, BERT does not "understand" everything in the text, it can not handle documents longer than 512 tokens at once, and has redundant components which remain unused in some tasks. Kovaleva and Romanov [18] found that BERT is highly overparameterized by disabling some parameters and then comparing to the initial results. For the above reasons, many optimized BERT models (e.g., ELECTRA, RoBERTa, DistilBERT, etc.) were released in the recent years. In this study, hoever, the original BERT is used, since it is still one of the most popular language models that demonstrates reliable results across many natural language processing tasks [1].

BERT Topic Modeling. Topic modeling is an unsupervised machine learning approach for detecting latent themes in a collection of texts based on clusters of words. One of the most used topic modeling methods is Latent Dirichlet Allocation (LDA). LDA identifies a set of words that are linked with each topic, as well as a set of topics characterizing each document in the corpus. As a main drawback of LDA, it builds on a probabilistic approach using the Bag-of-words model but ignoring the ordering and contextual meanings of the words [27]. Therefore, a context-aware BERT topic modeling approach [12] is used in this study to cluster words into meaningful groups. The processing steps of this method are shown in Fig. 3.

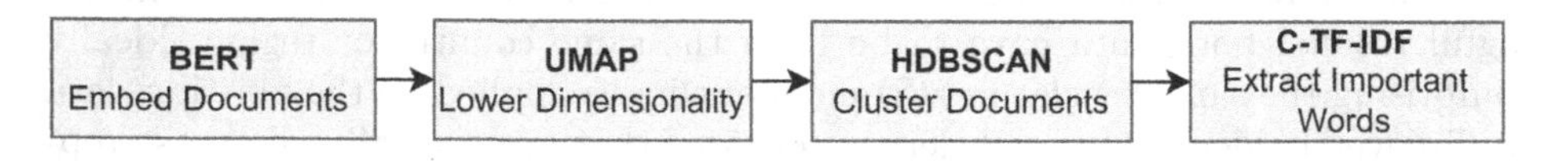

Fig. 3. An overview of the BERT topic modeling algorithm.

Dimensions of the petition data embeddings created with BERT are reduced using (Uniform Manifold Approximation and Projection for Dimension Reduction) UMAP [20] to lower the dimensionality of the word vectors while maintaining as much information as possible. After applying UMAP, semantically similar documents are clustered using (Hierarchical density-based spatial clustering of applications with noise) HDBSCAN. HDBSCAN is an extension of (Density-based spatial clustering of applications with noise) DBSCAN that groups data based on cluster stability: instead of looking for clusters of a specific form, it searches for data areas that are denser than the surrounding space. In contrast to, for example, k-means algorithm, HDBSCAN performs well even when the clusters have complicated shapes and a considerable amount of noise. In the next step, C-TF-IDF (Class-based Term Frequency-Inverse Document Frequency) scores are calculated for all the words to measure the importance of words within clusters. C-TF-IDF of term x within class c is determined as:

$$C - TF - IDF_{x,c} = tf_{x,c} \times \log(1 + \frac{A}{f_x}),$$

where $tf_{x,c}$ is the frequency of word x in cluster c, f_x is the frequency of word x across all clusters, and A is the average number of words per class. The higher the C-TF-IDF score, the more the word represents a particular cluster.

Once all words are extracted, the obtained topics are labeled by human annotators (the labels are not directly utilized in the proposed system, and the purpose of the annotation was only to help with human interpretation). The most representative petitions are retrieved from each topic using exemplar points from the HDBSCAN results to reduce the amount of data while keeping only the most important documents.

BERT Extractive Summarization. Representative petitions are summarized by applying BERT extractive summarization [22]. In contrast to abstractive summarization methods, extractive summarization does not generate any new text but outputs a summary made of the most relevant sentences of the original document. The algorithm is depicted in Fig. 4. The BERT extractive summarization method works by embedding sentences of the document, then running k-means to select sentences closest to the centroids. The number of sentences in the summaries is set, based on the length of the original input text.

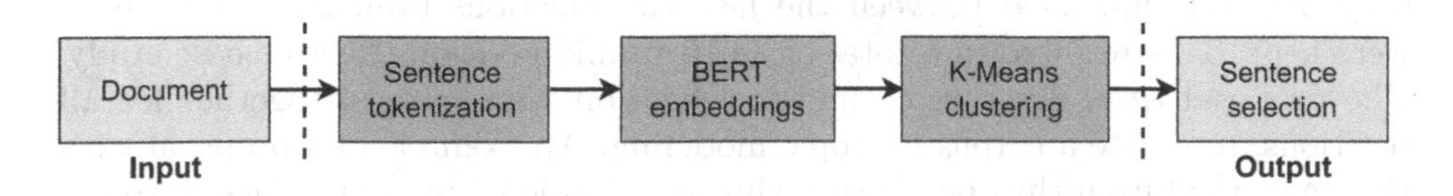

Fig. 4. Processing steps of the BERT extractive summarizer.

Similarity Calculation and Ranking. Semantic similarities between the contents of each VAA statement and summaries of the most representative petitions are measured by calculating the cosine similarity between the vectors in the interval of [0,1]. Since the original BERT model was not trained for sentence similarity task, and tends to perform poorly in this aspect [19], the so-called sentence-BERT model [24] is used in this study for sentence embedding, which produces fixed-sized sentence vectors that can be compared. After cosine similarity scores are calculated, the top five petition summaries with the highest scores are ranked for every VAA statement, resulting in the output suggestions.

4 Case Study

The proposed system can obviously be applied when both e-petitions and VAAs are part of the political landscape in a particular constituency. However, it can be assumed that this will increasingly be the case in the near future. The case analyzed in the paper is Taiwan that can be considered as the forefront of the developments in the fields of civic tech and digital democracy.

4.1 Data

The petition data used comes from the Taiwanese Join Platform that aims to reinforce communication between the citizens and the government by implementing an online petition system for policy proposals. While the platform contains other functionalities as well, the focus here is on e-petitions. The public administration formally checks the proposals in a "verification phase" before making them open for support votes. If the number of supporters reaches a minimum of 5,000 during the 60 day voting period, the government will provide an official response within a time frame of two months [15]. The proposal function is closed for 60 days prior to an upcoming election. On the national open data portal, the public administration updates an open dataset monthly about all proposals starting from 2015 September (https://data.gov.tw/dataset/58036), including information about the petitions, such as submission date, title, description, justification, number of supporters, topic category, and a unique identifier for each petition. For the purposes of this study, proposals that did not pass the verification stage or were withdrawn by the proposer were removed from the dataset. Furthermore, considering that issues of public concern may change over time, only petitions uploaded between the last two elections (Jun 2016–Nov 2019) were kept. This resulted in a total of 3,610 usable petitions for this case study. The proposal title, description, and justification were merged together for all petitions, to create a corpus for topic modeling. An example of a proposal with the items used from the open data is shown in Table 1 (due to the lack of space, only the English translations are shown here, the original is in Chinese).

The VAA data used in the form of policy statements originates from the 2020 presidential elections. The policy statements were manually retrieved from Taiwan's official VAA website, iVoter [16] (https://ivoter.tw). In total, there

Table 1. A translated example of an e-petition from the Join Platform.

Title	Legislation should prohibit dual citizenship
Description	1. Nationality should be associated with only one country 2. Nobody should be a citizen of both country A and country B 3. People who want to be naturalized must renounce their former citizenship
Justification	Our country should prohibit people from having dual citizenship, and from enjoying the resources of two countries (e.g., health care)

were 25 statements with short descriptions of the issues and position statements representing common opinions of the people. In addition, there are labels representing the topics covered in the statements. The labels for topics are *Domestic policy, Education, Labor, Environment, Economy, Cross-Strait relations, Judiciary, Diplomacy, National security, Constitutional government.* Table 2 presents a translated example from iVoter. In the similarity calculation step, only entries from the fields "Statement" and "Description" were used. Too much redundant information can affect the model performance of BERT.

4.2 Experiments

Experiments were conducted using a workstation with the following characteristics: NVIDIA Geforce RTX 2080 GPU, 128 GB RAM, and an Intel i9-9920X CPU. Considering the processing power of an average consumer single CPU as of 2022, computing BERT word embeddings without a discrete GPU would increase processing times substantially. In such cases, using cloud GPU/TPU platforms to reproduce results is recommended.

In the beginning, topic modeling using Chinese BERT with Whole Word Masking [6] was applied to the petition data corpus. The whole word masking approach optimizes the hidden word prediction task and tends to be more efficient for the Chinese language. The model was pre-trained on a large text collection, including Wikipedia, news data, encyclopedia, etc.

To choose the most representative topic model, preliminary experiments were conducted with different numbers of topics. Based on human assessment, a model with 50 topics was selected as it was found to be able to discover the most diverse but still not completely mutually exclusive topics. Finally, 20 words per topic with the highest C-TF-IDF scores were extracted for topic labeling. Topic annotation was conducted by the authors. To unify the possible topic labels, labels from the Level 2 theme codes of the Swiss Federal Office of Statistics (FOS) were used. The final label for each topic was decided through majority voting. Examples of the obtained topics are shown in Table 3. It should be noted

Table 2. A translated example of a VAA data point from iVoter.

Label	Diplomacy & National security
Statement	Do you support the idea of increasing the national defense budget?
Description	According to the 2019 Chinese government report, China's national defense budget is 1 trillion 189,876 billion (about NT$5,47 trillion); our total central government budget for FY108. This year's national defense budget is $340,478 billion, about 16 times difference between the two sides of the Taiwan Strait
Support	The military is for promoting peace. Considering the imbalance of military power across the Taiwan Strait, our country should increase its defense budget, promote national defense autonomy and purchase more military equipment
Against	Taiwan is a small country and is not suitable for an arms race with China. The military should be only defensive, and the budget saved should be allocated to education or social welfare

Table 3. Two examples of topics with their respective descriptions.

Words from BERT		FOS Topics	Words from BERT		FOS Topics
House	Raise		Plastic bag	Bag	
Housing price	Lease		Plastic	Plastic bag	
Householder	Investor		Straw	Reduce	
Rental	Room credit		Recycle	Packaging	
Housing	Price of land		Garbage	Green donation	
Empty house	Speculation	Housing	Pollution	Paper cup	Environment
Rents	Registration		One-time	Container	
Rents	Secondment		Environment	Pollution	
Rentals	Cost		Eco-bag	Carry bag	
Real estate	Market		Consumer	Resources	

that although synonyms may result in the same translation, all original topic words obtained are unique.

Selecting the most representative petitions for the topics with HDBSCAN reduced the initial number of 3,610 documents to just 216. Since BERT has a maximum length limit of 512 tokens for the input sentence, some petition sentences were split. The summaries' length was decided by the number of characters in the original text, using a ratio parameter. If the length of the document was more than 750 characters, the ratio of sentences to summarize was set to 0.3, meaning that 30% of the sentences, which are most representative, will be preserved. When the number of characters was in the range of 300 to 750, the ratio was set to 0.5. When the number of characters was less than 300, the document was not summarized to prevent losing important information from the originally short text (147 documents). Text summarization was applied to help VAA designers focus on the main idea of the text and save the time as well as to reduce the memory usage in the similarity calculation step.

Table 4. An example of the extractive petition summarization.

Original	English
提高公共政策網路參平台提案門檻嚴謹度剔除有心人士製造假議題造成。成案求精不求相政府需積極確實應提高門檻, 不能對造成造假議題有率的度僅通過電子郵件和官方文件無法解決。 不要使平台成渲染造成社立幼稚意形制造假工具浪行政社源。**1) 5000門檻太低數字。**何在感覺? **白宮願聯署門檻10萬人我建議提案門檻設為總人口數的千分之一2萬3千人或取整數2萬人。 2)認證機制太草率如兒戲杜絕殭屍假帳號重複投票附議提案附議應以正確身分證字號或護照編。**	**Raise the public policy online participation platform proposal threshold, block people with intention to create fake issues, which causes social conflicts and results in wasting resources.** The government needs to be active, the threshold should be raised, problems of fake news cases should not be taken lightly. This cannot be solved just by emails and official documents. Don't let the platform become a tool for rendering social opposition, childish ideologies, creating fake issues, and wasting administrative and social resources. **1) The threshold of 5000 is too low.** What do you think? **The White House wants to establish a threshold of 100,000 people. I suggest that threshold should be 23,000 or a round number, e.g., 20,000. 2) There are too many loopholes in the authentication mechanism, it's like a child's play. It is necessary to prevent re-voting by zombie and fake accounts. When submitting a proposal and verifying it, you should need to enter your ID number or upload your photo.**

Table 4 presents an example of a summarized petition. On the left, there is the original text written in traditional Chinese, and the English translation is on the right. The particular petition addresses the problem of manipulating e-participation platforms. Bold text represents the extractive summary generated by BERT, where the main idea, including the claims and suggestions, is preserved.

Finally, the semantic similarity between summaries and the 25 VAA policy statements was calculated. The petition summaries were ranked for every VAA statement, based on their similarity score to find what documents are topically relevant to the VAA items. In addition, to be able to address the importance of the highly-ranked petition summaries, the number of votes the petitions received (as an indicator of the petition saliency) is preserved for the final output of VAA statement suggestions.

5 Results and Discussion

The proposed system was calibrated to generate the top five petition summaries with the highest similarity scores. Organized by VAA statement, five suggestions is a quantity a human VAA designer could easily digest and review on one page of paper or fit on a computer screen for reviewing. Due to the lack of space, only two system suggestions are listed here (see Table 5). All the suggestions the system produced, including the labeled topics with their respective descriptions, are from: https://github.com/DBurya/vaa-redesign-suggestions.

Table 5. Example of two system suggestions for one corresponding VAA statement.

VAA statement
Do you agree that Taiwan should completely abandon the use of nuclear energy?
System suggestion 1
Nuclear power plants in Taiwan must be used to generate electricity. Use one or three nuclear power plants to generate electricity. Three nuclear power plants will definitely be able to generate tens of millions of watts of electricity. It is highly recommended to use nuclear power plants in Taiwan. We must immediately use nuclear fusion to generate electricity **Petition summary similarity score:** 0.808 **Number of votes:** 9
System suggestion 2
The government has spent $ 8 billion to buy nuclear fuel rods, and over the last three years sold them back to the US at a low price of $ 2 billion ($600 billion/year). Taiwan may need to spend more than 10 billion dollars if government wants to repurchase these fuel rods in the future. 2) Taiwan is already suffering from a lack of electricity, and will have to build a coal-fired power plant with high-levels of pollution, and press the start button on the two units of Lungmen Nuclear Power Plant. In the future, if Taiwan wants to import fuel, this must be done with the assent of the US. The US may not be interested in selling fuel rods to Taiwan, so Taiwan's energy autonomy will be weakened. Moreover, if China surrounds Taiwan, Taiwan will have no electricity available in 30 days **Petition summary similarity score:** 0.660 **Number of votes:** 5956

In a VAA, besides actual policy statements, the most common opinions in favor or against are usually also included to help the users form an opinion. However, due to lack of space they are omitted here. Table 5 shows the example of two system suggestions for one such VAA policy statement. The statement refers to the controversial political discussion whether Taiwan should completely abandon the use of nuclear energy or not. According to the World Nuclear Association (2021), nuclear power generates 12 percent of Taiwan's total electricity [5]. In 2016, the government has declared its willingness to make Taiwan free from nuclear energy by 2025. However, in a 2018 referendum vote, 59 percent of the people voted against the concept of a nuclear-free state [29].

The petition summary similarity score of 0.808 for system suggestion 1 indicates a high similarity with the text in the respective VAA statement. However, during the 60 days this e-petition text was available on the Join platform to collect support votes, only nine supporters endorsed the proposal. While the similarity score of 0.66 for system suggestion 2 is considerably lower, the number of endorsing votes indicates a high popularity among residents. That particular e-petition even surpassed the threshold of 5,000 votes and thus triggered an official response by the public administration unit in charge. The discrepancy between the similarity score and the number of votes highlights their complementary and that taking both criteria into account will be the most beneficial approach for VAA designers.

The content of both system suggestions coincides in many respects with the original VAA policy statement. However, suggestion 2 clearly brings up content which neither the VAA statement nor the system suggestion 1 are referring to. In addition to the original statement, issues such as the cost, reverting back to coal as well as dependencies to the USA or China are brought up. VAA designers therefore might completely revise or at least consider a reformulation of the original policy statement.

As a practical implication, the proposed system and approach for VAA statement selection and design leads to a re-emphasis of the core of the voting advice application as a civic tech tool, the policy statements themselves. In the past, little attention was given to the question of what constitutes the source and the criteria for statement selection. Although oftentimes loosely based on news and party manifestos, the task of statement selection is still largely considered to be the work of the ones in the know, the experts. However, even for experts it is very difficult to impossible to know the universe of potentially well suited VAA policy statements. Whereas the proposed system still needs human intervention, it nevertheless adds a much larger and known pool thereof, making the process of statement selection more transparent. In the best case, the increased availability of open government data such as the access to e-petition data might lead to the need of VAA designers to open up the black box of statement selection and document the process.

6 Conclusions

In this work, a system was proposed to assist the design of VAAs by generating statement-wise suggestions for VAA designers from open government e-petition data. The most representative documents were retrieved from petition data using BERT topic modeling. The documents were summarized to ease the burden on VAA designers in terms of manual assessment of the data. In the end, summarized documents were compared with the VAA data to produce the most relevant VAA statement suggestions. A case study was performed on Taiwanese VAA (iVoter) and e-petition data (Join Platform), with promising results allowing for the revision of the original or the creation of new VAA policy statements. The paper adds novelty to the domain of study by providing a reproducible support solution for the design of a VAA on one of the costly parts, mainly performed by humans so far. In many VAAs, the empirical basis for the policy statements being used is not empirically grounded.

While the primary purpose of the suggested system is to reduce the cost of VAA statement design, there are additional use case scenarios. Since the quality of a VAA largely depends on the policy statements being used, testing those statements beforehand or in the early stages when going online is of utmost importance. However, proper testing requires the availability of a whole series of alternative policy statements across the whole spectrum of political issues under concern before election date. The proposed system would allow for exactly that and offer a variety of policy statements readily available for testing.

Besides the technical limitations of BERT, there are obvious practical concerns to consider. In case a constituency does not have or just started to use e-petitions there is simply no data available to run the proposed system. The same holds for VAAs. The assumption is to have both data sources readily available in an open format accessible to the researchers. A further underlying assumption is that e-petition texts collected for a legislature leading up to an election would provide for a valid basis to formulate VAA policy statements. At the moment, there is no validation or proof given. However, an argument in favor of texts stemming from e-petitions is that they are usually broad in scope, provide for the right amount of text to apply BERT and are political by definition.

This research is the first of its kind, in which petition data and the BERT language model are used to assist VAA designers in creating policy statements. It provides for an illustration of how open government data can help to build up a basis and ecosystem for civic tech tools, in this case open e-petition data feeding into the design of VAA policy statements. In future work, it is planned to expand the proposed system on different data sources as input but also to different constituencies. In addition, a step of text segmentation such as in [14] could be added prior to the proposed system. Such an extended version would allow for the inclusion of different political text sources, most notably party manifestos. With some customization, the system can also be used to generate completely new VAA statements automatically, thus further facilitating the work of VAA designers.

References

1. Acheampong, F.A., Nunoo-Mensah, H., Chen, W.: Transformer models for text-based emotion detection: a review of BERT-based approaches. Artif. Intell. Rev. **54**(8), 5789–5829 (2021). https://doi.org/10.1007/s10462-021-09958-2
2. Anderson, J., et al.: Matching Voters with Parties and Candidates: Voting Advice Applications in Comparative Perspective. ECPR Press, Colchester (2014)
3. Anwar, A., Ilyas, H., Yaqub, U., Zaman, S.: Analyzing QAnon on Twitter in context of US elections 2020: analysis of user messages and profiles using VADER and BERT topic modeling. In: DG.O2021: The 22nd Annual International Conference on Digital Government Research, pp. 82–88. DG.O 2021, ACM, New York, NY, USA (2021). https://doi.org/10.1145/3463677.3463718
4. Arana-Catania, M., et al. : Citizen participation and machine learning for a better democracy. Digit. Gov.: Res. Pract. **2**(3) (2021). https://doi.org/10.1145/3452118, https://doi.org/10.1145/3452118
5. World Nuclear Association: Nuclear Power in Taiwan. https://world-nuclear.org/information-library/country-profiles/others/nuclear-power-in-taiwan.aspx (2021). Accessed 12 Dec 2021
6. Cui, Y., et al.: Pre-training with whole word masking for Chinese BERT. arXiv pp. 1–11 (2019). https://arxiv.org/pdf/1906.08101.pdf
7. Devlin, J., Chang, M.W., Lee, K., Toutanova, K.: BERT: pre-training of deep bidirectional transformers for language understanding. In: Proceedings of the 2019 Conference of the North American Chapter of the Association for Computational Linguistics: Human Language Technologies, vol. 1 (Long and Short Papers), pp. 4171–4186. Association for Computational Linguistics, Minneapolis, Minnesota, June 2019. https://doi.org/10.18653/v1/N19-1423
8. Gaglani, J., Gandhi, Y., Gogate, S., Halbe, A.: Unsupervised WhatsApp fake news detection using semantic search. In: 2020 4th International Conference on Intelligent Computing and Control Systems (ICICCS), pp. 285–289 (2020). https://doi.org/10.1109/ICICCS48265.2020.9120902
9. Garzia, D., Marschall, S.: Voting advice applications under review: the state of research. Int. J. Electr. Govern. **5**(3–4), 203–222 (2012)
10. Gemenis, K.: An iterative expert survey approach for estimating parties' policy positions. Qual. Quant. Int. J. Methodol. **49**(6), 2291–2306 (2015)
11. Giachanou, A., Zhang, G., Rosso, P.: Multimodal multi-image fake news detection. In: 2020 IEEE 7th International Conference on Data Science and Advanced Analytics (DSAA), pp. 647–654 (2020). https://doi.org/10.1109/DSAA49011.2020.00091
12. Grootendorst, M.: Bertopic: neural topic modeling with a class-based TF-IDF procedure, pp. 1–10. arXiv preprint arXiv:2203.05794 (2022)
13. Hagen, L.: Content analysis of e-petitions with topic modeling: how to train and evaluate LDA models? Inf. Proces. Manag. **54**(6), 1292–1307 (2018). https://doi.org/10.1016/j.ipm.2018.05.006
14. Hananto, V.R., Serdült, U., Kryssanov, V.: A text segmentation approach for automated annotation of online customer reviews, based on topic modeling. Appl. Sci. **12**(7) (2022). https://doi.org/10.3390/app12073412
15. Huang, H.Y., Kovacs, M., Kryssanov, V., Serdült, U.: Towards a model of online petition signing dynamics on the join platform in Taiwan. In: 2021 Eighth International Conference on eDemocracy eGovernment (ICEDEG), pp. 199–204 (2021). https://doi.org/10.1109/ICEDEG52154.2021.9530852

16. iVoter: Taiwan's voting advice application. http://ivoter.tw/ (2013). Accessed 10 Oct 2021
17. Katakis, I., Tsapatsoulis, N., Mendez, F., Triga, V., Djouvas, C.: Social voting advice applications-definitions, challenges, datasets and evaluation. IEEE Trans. Cybern. **44**(7), 1039–1052 (2013)
18. Kovaleva, O., Romanov, A., Rogers, A., Rumshisky, A.: Revealing the dark secrets of BERT. In: Proceedings of the 2019 Conference on EMNLP-IJCNLP, pp. 4365–4374. Association for Computational Linguistics, Hong Kong, China, November 2019. https://doi.org/10.18653/v1/D19-1445
19. Li, B., Zhou, H., He, J., Wang, M., Yang, Y., Li, L.: On the sentence embeddings from pre-trained language models. In: Webber, B., Cohn, T., He, Y., Liu, Y. (eds.) Proceedings of the 2020 Conference on Empirical Methods in Natural Language Processing, EMNLP 2020, Online, 16–20 November 2020, pp. 9119–9130. Association for Computational Linguistics (2020). https://www.aclweb.org/anthology/2020.emnlp-main.733/
20. McInnes, L., Healy, J., Melville, J.: UMAP: uniform manifold approximation and projection for dimension reduction (2018). https://arxiv.org/abs/1802.03426
21. Mendez, F.: Modeling proximity and directional decisional logic: what can we learn from applying statistical learning techniques to VAA-generated data? J. Elect. Public Opin. Parties **27**(1), 31–55 (2017)
22. Miller, D.: Leveraging BERT for extractive text summarization on lectures (2019). https://arxiv.org/ftp/arxiv/papers/1906/1906.04165.pdf
23. Reiljan, A., da Silva, F.F., Cicchi, L., Garzia, D., Trechsel, A.H.: Longitudinal dataset of political issue-positions of 411 parties across 28 European countries (2009–2019) from voting advice applications EU profiler and euandi. Data Brief **31**, 1–9 (2020). https://doi.org/10.1016/j.dib.2020.105968
24. Reimers, N., Gurevych, I.: Sentence-BERT: sentence embeddings using Siamese BERT-Networks. CoRR pp. 3982–3992 (2019). http://arxiv.org/abs/1908.10084
25. Shirafuji, D., Kameya, H., Rzepka, R., Araki, K.: Summarizing utterances from japanese assembly minutes using political sentence-BERT-based method for QA Lab-PoliInfo-2 Task of NTCIR-15. CoRR, pp. 1–8 (2020). https://arxiv.org/abs/2010.12077
26. Silva, N., et al.: Evaluating topic models in Portuguese political comments about bills from brazil's chamber of deputies. In: Britto, André, Valdivia Delgado, Karina (eds.) BRACIS 2021. LNCS (LNAI), vol. 13074, pp. 104–120. Springer, Cham (2021). https://doi.org/10.1007/978-3-030-91699-2_8
27. Silveira, R., Fernandes, C.G., Neto, J.A.M., Furtado, V., Filho, J.E.P.: Topic modelling of legal documents via LEGAL-BERT. In: RELATED 2021, Relations in the LegalDomain Workshop, in conjunction with ICAIL. pp. 64–72. CEUR-WS.org, Online (2021). http://ceur-ws.org/Vol-2896/
28. Terán, L., Mancera, J.: Dynamic profiles using sentiment analysis and Twitter data for voting advice applications. Gov. Inf. Q. **36**(3), 520–535 (2019). https://doi.org/10.1016/j.giq.2019.03.003
29. World Nuclear News: taiwanese vote to keep nuclear in energy mix. https://www.world-nuclear-news.org/Articles/Taiwanese-vote-to-keep-nuclear-in-energy-mix (2021). Accessed 12 Dec 2021

Investigating Trust and Risk Perceptions in a Hybrid Citizen Journey

Christian Gutowski(✉) and Jing-Heng Kao(✉)

Department of Information Systems, Westfälische Wilhelms-Universität Münster, Leonardo-Campus 3, 48149 Münster, Germany
{ch.gutowski,jkao}@uni-muenster.de

Abstract. Key public services are increasingly offered digitally due to different government policies. However, many administrations are still in a transition phase from traditional to digital services and the use rates by citizens present a concern. Trust is one crucial factor for adoption. In this paper, we approach the topic of trusting beliefs in relation to risk perceptions and along specific steps of the service delivery process in a hybrid service delivery setting. For this, we utilized a citizen journey as an analytical tool and conducted a qualitative interview study. Two hybrid public services in Germany, residency registration at a city administration and the enrolment at a public university, were investigated with national and international university students. This study delivers a proof of concept for the citizen journey to be used as a participatory analytical tool that can gain insights related to the design of public service delivery. Our results show that trust and risk perceptions are relevant for the citizens' impressions of public services. Even though the interviewees perceived risks regarding the digital abilities of the administrations to provide secure digital services, they would still prefer to conduct them entirely online. This is due to existing convenience and communication issues, but also due to trusting beliefs in the goodwill and integrity of local German administrations. With this, we contribute to current debates on the citizen-orientation of public service delivery.

Keywords: Qualitative research · Trust · Risk perception · Citizen journey

1 Introduction

In 2021, the European Commission published a digitalization compass [10], detailing a vision for digitalization in this decade including all key public services online availability. However, according to the 2021 e-government benchmark by the European Commission [13], the overall e-government maturity of European countries varies notably. While many countries, especially the Nordic and Baltic ones, have achieved substantial digitalization of public administrations, the big economies and south-eastern countries lag behind [13]. In many countries, a

Published by Springer Nature Switzerland AG 2022
R. Krimmer et al. (Eds.): ePart 2022, LNCS 13392, pp. 35–51, 2022.
https://doi.org/10.1007/978-3-031-23213-8_3

transition phase exists between the traditional (analogue) and purely digital government [2]. While the digital channels may already be present, many citizens combine or avoid using them. Therefore, discussing the factors that influence their channel choice in a hybrid services environment is important [20]. Achieving the stated goals requires not only the extensive provision of digital public services but also citizens using them - an issue with a long-standing tradition in both practice and research. Contrasting current use statistics [13,35] with the ambitious goals set forth by the European Commission, the issue of citizens' use behavior still occupies researchers and practitioners. To increase the adoption of digital public services, citizen-orientation and active citizen-participation in service innovation and design are promoted as means to raise use rates.

Germany is an interesting case as considerable efforts have been made to increase both citizen-orientation in service design and, subsequently, use rates. The German online access act, put in force 2017, obligates administrations of all hierarchical levels to make their public services accessible online [15]. Although considerable funding of more than 3.3 billion euro has been made available for implementation, the current outcomes fall well short of expectations and targets [26]. In spite of these significant investments and reduced contacts during the COVID-19 pandemic, Germany consistently underperforms in digitalization and key enablers [14]. Furthermore, the adoption rate stagnates with only 52% of citizens using digital public services and the share of citizens satisfied with their availability on a local level has even fallen from 62% to 47% in 2021 [21]. Due to the German administrative structure, most of the policy implementation and public service delivery lies with local authorities [33] with 80% of public services delivered by municipalities [19]. Half of the provided funding is therefore received by lower level administrations [26] and the online access act builds on measures of co-creation with citizens and measures its success in the rate of acceptance and use of them [15]. Thus, how citizens perceive their local administrations and public services is important since the overall success of digital transformation in the German public sector rests on the shoulders of municipalities.

Repeatedly, scholars have highlighted the importance of trust for citizens' use of public services - via digital means but also on-site. From the citizens' perspective, trust in the context of e-government may be influenced notably by their views and beliefs [4]. Trust can increase the usage of digital public services as it can make citizens more comfortable to use them [3] and in Germany it is a factor for low acceptance of e-government services [1]. Trust and risk perceptions can be seen as mirror images [9]. Depending on how prevalent either or both perceptions are, individuals engage in trusting behavior or risk taking. Better understanding trust beliefs and risk perceptions of citizens in relation to public services can help administrations to comprehend citizens' choices and increase adaption as it is one of the major goals by the European Union and its members.

Our study selects university students as exemplary citizens, assuming that their perspective can contribute to innovating public services. The European Union aims to increase and ease student mobility both within and outside the EU, e.g., via Erasmus+ [12] or the Bologna process [5], both contribute to the annual increase of international students in Germany [17]. Because public administration is transforming to become a more citizen-oriented service provider [1], more international students play an increasing role as service consumers in this sector. Interviewed students are from Germany as well as other selected countries, in order to discuss trust beliefs and risk perceptions between citizens with different backgrounds and the German public administration.

Thus, the research goal of this paper is to identify trust and risk perceptions in the citizen-administration-relationship in different phases of hybrid public services. For this, a citizen journey is used as analytical tool and local and state administrative services for national and international university students in Germany are used as an example. This study applies the citizen journey proposed by Scholta et al. [29], tailored to the relationship between citizens and the public administration as public service delivery transforms from supplier- to customer-orientation [22]. We adapt the generic citizen journey based on two hybrid public services at the local level, residency registration at a German city administration and enrolment at a public German university, and conduct a qualitative interview study. The remainder is structured as follows: First, we introduce the basic concepts of trust and generic citizen journey. Next, we explain the methodology of service selection, tool adaption, interview conduction and analysis process. Subsequently, we present our findings according to factors of perceived trustworthiness, before discussing trust and risk perceptions and concluding the paper.

2 Research Background

Research discussing trust and public administration already exists [3] but the debate of trust-related issues regarding technological developments persists [11]. The existing literature is mainly concerned with the generic level of trust [1] but does not concentrate on the single steps within one service. Different citizen groups may have different experiences and opinions regarding trust and risk-related issues. This paper select students as the target group to account for the increase in students in Germany, and the integral role that the internet plays in their daily lives [2]. Next, in-depth exploration of trust and risk perceptions within specific services is crucial, in order for researchers to fully grasp the variety of trust perceptions in the public sector and further develop the current knowledge and literature in that area. Furthermore, a better understanding of trust concerns between citizens and local-level public administrations can contribute to service providers optimizing services and digitalization in the future.

2.1 Trust and E-Government

Researchers have described trust as a multidisciplinary concept used in many fields [8]. In this study, we apply one extensive definition proposed by Mayer et al. [24] which defines trust as "the willingness of a party to be vulnerable to the actions of another party based on the expectation that the other will perform a particular action important to the trustor, irrespective of the ability to monitor or control that other party" ([24], p. 712). Mayer et al. [24] distinguish three trusting beliefs that - together with the trustor's propensity to trust and risk perceptions - impact trust and trusting behavior, i.e., the trustees perceived benevolence (or goodwill), integrity, and competence. Here, trust is a decision heuristic that enables people to act in certain situations that are perceived risky (e.g., regarding a loss of money or privacy). Those risks can be perceived differently by individuals who need a corresponding coping mechanism. Hence, trusting beliefs influence the risk perceptions and vice versa [9]. Trusting beliefs also influence the individuals' strategies on how to handle perceived risks.

Trust is relevant for public administrations as It can make citizens comfortable (or not) to use public services - independent of the mediating channel - provided by the government [3]. In digital settings in particular, trust becomes a central element for citizens' use of public services as current studies indicate. For example, Akkaya et al. [1] discussed trust as crucial in general, with culture playing a role that may impact citizens' trust to e-government. Distel et al. [11] summarized a general model of citizens' trust in the context of e-government, distinguishing the technical infrastructure and the public administrations as potential trustees. Some general-level research applies analytical frameworks to assess the interplay between digital technologies and trust in the context of e-government [4,25,31]. Other research focuses on citizens' e-government trust in certain areas [7,27]. In addition, Cabinakova et al. [27] studied cultural trust and habit patterns that may influence the citizens' adoption of e-Government. These papers did not study different steps of specific administrative services with a particular governmental level but concentrated on the relationship between digital transformation and trust, whereas this paper discusses two specific administration services along the complete delivery process and focuses on one group by identifying different nationalities living and studying in Germany. Against the background of many countries facing transformation from traditional public service delivery into digitalized public service delivery [2], it is important to build an understanding of whether citizens perceive differences between traditional and digital service delivery within one and the same service.

Considering literature on e-government and trust research, the following factors of perceived trustworthiness are important to consider when studying trust and risk perceptions in relation to (digital) public services: perceived goodwill and values, perceived competency, perceived data security and protection, information, communication, convenience, and personal preferences (see Table 1).

2.2 Citizen Journeys as Analytical Tool

We utilized a tool called the citizen journey to analyse and evaluate interviewees' trust beliefs and risk perceptions in two transactional services. With the growing number of touchpoints in different channels and technologies in the past decade, more awareness for the customer's perspective is noted, attracting more attention to firms that deliver and optimize their services using the so-called customer journey as a conceptualized tool for customer-company relationships [23]. The principles of the customer journey can be transferred to the field of public administration, where significant attention is paid to increasing transparency, efficiency and customer orientation [1]. As many countries are making the transition from analogue to digital public services [13], a combination of both is common. Understanding where citizens differentiate between the two within the same service is important. This increases the importance of an in-depth exploration of the citizen perspective on the relationship to public administration services, with the goal of providing the public sector with useful insights and encouraging transformation. A generic citizen journey that helps researchers and practitioners to better understand and design the public service delivery process from a citizen perspective was proposed by Scholta et al. [29]. In a recent study, Graf-Drasch et al. [18] utilized and re-defined this citizen journey to build an extensive understanding of a citizen in the context of smart and sustainable districts. The citizen journey visualizes the delivery of a public service step by step. Therefore, practitioners can use it to design and optimize service processes and researchers to investigate citizen experiences [29]. In this paper, we build up on the prior research by using the citizen journey as analysis tool, adapting it to conduct research in terms of two specific services. Thereby, it enables us to add insights about factors that may impact citizens' experiences and trust concerns. This can contribute to future optimization regarding digital public services.

3 Methodology

Public Services Selection. This research takes the citizen perspective and therefore set the following requirements for the service selection before conducting the interviews: First, as 80% of public services in Germany are provided on a local level [19], we chose to focus on services provided within the city of interest. Second, the services must be provided to German, EU and non-EU citizens alike. After an analysis, we selected the residence registration at the city administration and the enrolment at the university as it is a public (state) institution. Both public services are hybrid and therefore contain online and offline components.

Generic Citizen Journey Adaption. We adapted the generic version of the citizen journey by Scholta et al. [29] to the two services in question. As an example, one adapted citizen journey used in our study is visualized in Fig. 1. The three phases of every citizen journey are as follows: preparation, application, and results. The optional steps are written in italic, the other steps are mandatory.

Residency registration is mandatory for every citizen living in Germany for more than 3 months. First, in the preparation phase citizens are asked to provide their data and residential address to the local registration authorities. Therefore, citizens are forced by law to use this service [16]. After recognize this need, they either make an appointment through the city's online portal or go directly to the city hall for a spontaneous appointment. In the first case, the citizens receive an activation and subsequently a confirmation email. Citizens have the option to search for more information themselves, and the appointment can also be cancelled. Thus, already in the first step, citizens can decide whether they would like to conduct the service (at least partially) online or (completely) offline. Secondly, in the application phase, citizens prepare a valid identification and a certificate issued by the residency provider (e.g., landlord) and visit the city hall in person to carry out the registration. Currently, all steps in this phase are only offered offline, i.e., citizens cannot make a choice here. Thirdly, in the result phase, citizens may ask additional questions to the administrators, or the application may fail, or citizens may be asked to provide further documents later. After successfully completing all steps, citizens receive the result instantly and a letter with a tax number is sent afterwards, which is only provided offline as well. Finally, citizens can use the result for work or further administrative services.

The second service, enrolment at the university, is hybrid too. First, in the preparation phase, the need is triggered by receiving an admission letter. Students receiving a digital letter of admission must either accept or decline the offer in the university's online portal before the deadline. The decision may fail because students reject the offer or miss the deadline. Only upon accepting the study offer, students receive a digital enrolment form with detailed enrolment information. They then have the opportunity to ask questions and search information about enrolment online (e.g., website, e-mail) or offline (e.g., telephone, visiting the office). Secondly, in the application phase, students are required to prepare the relevant application documents (e.g., high school/bachelor diploma, transcripts of records), which they either have to send by post or hand in personally. The submission of forms and documents cannot be conducted online. Thirdly, in the result phase, students receive the enrolment confirmation and student ID-card by email and post. However, they may be asked to submit additional documents, or the enrolment may fail because the application was not received on time, post loss, etc. As a result, students can then pay the semester fees and use further services, such as access to reduced food, cultural events, access to university sites, and - most importantly - access to study programs.

Interviews. The relationship between e-government and citizens can be explored from the citizen's perspective through qualitative research, i.e., personal interviews. This is the most suitable research method for in-depth exploration of a personal topic such as trusting beliefs and risk perceptions [30]. Therefore, we have conducted six one-on-one interviews in this study in order to analyze public service delivery from the citizens' perspective. With student mobility in mind, we chose German, EU and non-EU citizens as our interviewees. Then, we selected interviewees within the authors' personal networks who fulfilled the nationality criteria and are not all enrolled in the same field of study to ensure a more diverse point of view. We used semi-structured interviews to gain a better understanding of interviewees' experiences with special consideration of factors of perceived trustworthiness (see Table 1). All interviewees are currently enrolled at the investigated university and did these two services. In Table 2 the demographic information of the interviewees is summed up.

Transcription and Analysis. After completing the interview phase, we transcribed all recordings. Next, we used MAXQDA2022 to code and do further analysis. We mainly utilized a deductive approach relying on existing trust-related

Table 1. Factors of perceived trustworthiness

Category	Description	Exemplary questions
Perceived different steps of citizen journey	Based on the trustor's (citizen) past experience, whether he/she has a different order for processing the two services compared to the revised citizen journey	Does the journey cover your experience? If not, which steps did you not need or are missing? Which steps in the journey did you perceive differently?
Perceived goodwill and values [4,24,28]	It describes how trustees (public administration) are believed to be willing (goodwill and integrity) to provide a good service for all citizens and add value	How did you perceive the integrity and goodwill of the administration? Do you believe in general that the administration shares the same values and interests and is willing to provide good online services?.
Perceived competency [4,28]	From trustor's perspective, are trustees (public administration) qualified and do they have sufficient skills to influence/do their job (ability) for trustors (citizen)	How did you perceive the competence of the administration?
Perceived data security and protection [4,8,11,28]	It refers to how trustors (citizens) perceive all risks that relate to the technical security of application systems	What do you think about the ability of the administration to provide data security/protection?
Information [4,8,28,31]	It describes how trustors (citizens) perceive the quality of information they can receive from trustees (public administration), including complete, comprehensible, correct information	Were there any obstacles or barriers during the process? e.g.,incorrect/incomplete/ incomprehensible information
Communication [28]	Trustors (citizens) perceive all the risks that relate to communication, such as contact through different channels (e.g., telephone, e-mail etc.) and misunderstandings caused by language barriers	Were there any obstacles or barriers during the process? e.g., obstacles during communication such as language barriers or misinterpretations
Convenience [4,28]	It describes how convenient the trustors (citizens) consider the service process to be, including time spent, ease of use, issue with post-delivery providers	How convenient do you think was the process, e.g., regarding the time spent?
Personal preferences	It describes the individual preferences of citizens in service delivery process, such as the impact of the pandemic to original lifestyle, personal character traits etc.	When and why did you choose or switch channels?

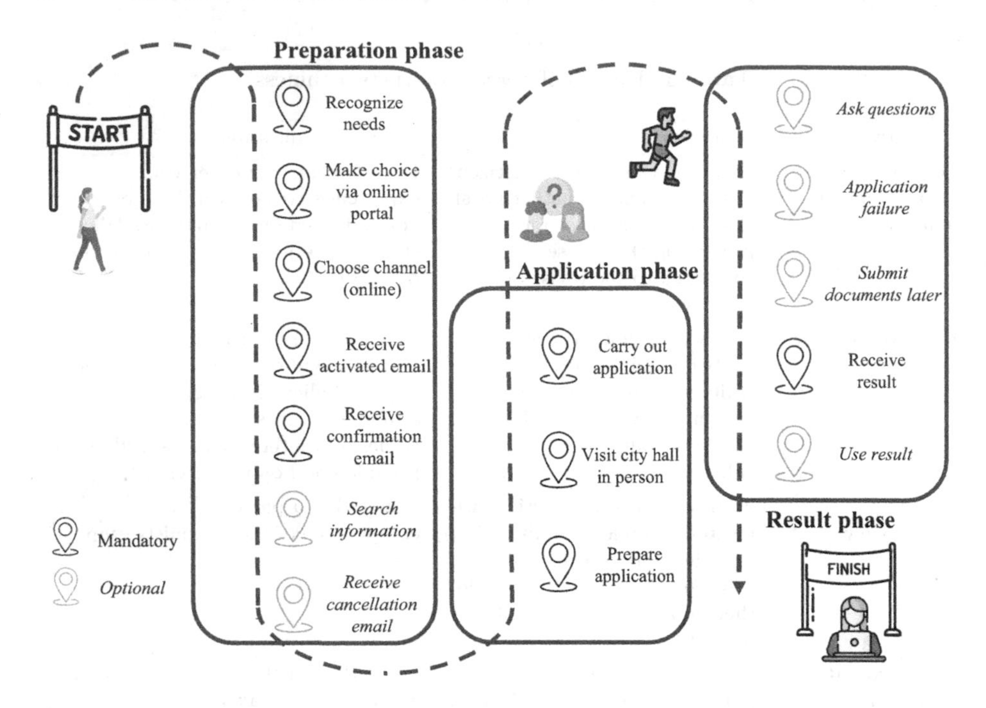

Fig. 1. Citizen journey of residency registration

Table 2. Summary of interviewees

Index	Nationality	Gender	Age	Field of study
A	Germany	Male	23	Computer Science
B	Germany	Male	24	Information Systems
C	Hungary	Female	24	Information Systems
D	Bulgaria	Female	25	Business Administration
E	China	Male	22	Business Administration
F	Taiwan	Female	26	Pharmaceutical Science

categories to code our transcription [32]. As listed in Table 3, eight categories were used for the coding phase to capture factors related to trust and public administration. Six categories were derived from existing literature about influence factors regarding the trust level of citizens, for example the perceived competency of an administration [4,24,28]. Furthermore, Yang [34] argues that the attitude towards trust is based on citizens' prior experience rather than present ones. Hence, we added the category of perceived different steps to discuss interviewees' experiences from the past. During the analysis, we found that personal preferences and character traits also play a significant role and therefore added this as an additional category. The coding and reviewing was done by one author at first and was then reviewed by the second author. Afterwards, the results were discussed. Our findings are presented and discussed in the next sections.

Table 3. Coding system categories

Category	Exemplary quote from the interviews
Perceived different steps of citizen journey	"Because it was during Covid-19, and they operated it only a few hours a day. I think you were not able to book an appointment at that time" (Interview C)
Perceived goodwill and values [4,24,28]	"I think people working in governmental institutions, in general, especially on the local level, maybe in [city name] rather than on the national level, have the best interests of all of us in mind. I'm quite certain of that, yeah" (Interview B)
Perceived competency [4,28]	"The quality of your work is not always that good, and they do not have the money for it. Like if you need somebody who make like a very good infrastructure for you. You need to pay them a lot of money and with them, the job opportunities in the public sector, it is not doable with this kind of salary"(Interview A)
Perceived data security and protection [4,8,11,28]	"I'm very trusting in that regard (data security and protection). I mean, I really don't have any fears regarding that. I feel like if a program like this is put into motion, then obviously there's going to be a lot of people who would need to check if everything's okay and everything is safe and secure" & "I always thought that my data was secure"(Interview D)
Information [4,8,28,31]	"Yes but I would not say something is missing, maybe it was not that clear or was not well formulated. I do not know if I got all the information from that email, or I needed to search something online, but I know that after that email it was definitely not 100 % clear. I did not have all the information at one place." (Interview C)/"It depends on who pick up the call, some people working at the university at the registration and enrolment office, I do not want to be mean, but sometimes they do not tell you everything, they just tell you well it is against of policy for privacy, we cannot tell you your documents are already here or not, we will tell you when you are enrolled or rejected" (Interview E)
Communication [28]	"But I prefer to do this stuff in person. Because I think there is less misunderstanding, and I think it is especially since I am a foreigner and I do not speak the language that well. I think I can avoid some misunderstanding better if I am going." (Interview C)/"Because I think then I have to speak in German, and I am not so confident with my German. The language is a barrier for me" (Interview F)
Convenience [4,28]	"Well along with the university, I do not think so but with post office yes. Because I wanted to track the shipping, but sometimes, they do a really bad job with it. They do not give you any information. They just show it is on the way. You never know when it is gonna arrive." (Interview E)/"Yes, but express delivery, and sometimes if there is a failure, like lost documents, then I will have to do it again" (Interview F)
Personal preferences	"I would say mostly positive. As most people in our age, I kind of grew up as technology was changing vastly." (Interview D)/"Now, since I am so depressed because of Covid-19, I will definitely do it in person just to have the feeling back. I would really appreciate of waiting in the office's waiting room" (Interview E)

4 Findings

Overall, six interviews were conducted with a similar length of between 46 and 66 min and averaging 55 min. During their studies the interviewees naturally made more experience with the university than with the city administration, and those prior experiences particularly influenced the perceived digital competency. Furthermore, while their own technological abilities varied among the interviewees, they generally had a positive attitude towards technology. The risks of digitalization such as data privacy are known and were mentioned but were not perceived as a major downside. Connected to the tech-savviness of the interviewees is the found preference for digital contacts over contacts in-person or via phone by all but one. Henceforth, the results will be presented based on the factors of perceived trustworthiness described in Table 3. The interviewees were asked about them, therefore, to give a more genuine impression of the interviewees' opinions, we did not combine these into the general notion of trust but present and discuss our findings based on the categories used in our interviews.

Perceived Different Steps of Citizen Journey. The interviewees mostly confirmed that the two presented citizen journeys covered their experience accurately. Seven out of twelve citizen journey experiences were confirmed completely with the other five only marginally differing - e.g., the registration at the city administration without an appointment or a differently perceived application form. The final revised citizen journey of residency registration as shown to the interviewees is illustrated in Fig. 1.

Perceived Goodwill and Values. The city as well as the university were in general perceived in a positive light. It was mentioned that the interviewees have been treated fairly and respectfully but that it is difficult evaluating the administrations' values as the standardized processes would not allow for that. However, a positive attitude especially towards governmental institutions on a local level was noted as those will "have the best interests of all of us in mind" (Interview B). Furthermore, it was mentioned that efficient services are a common interest of both the administrations and the students. However, the human aspect was mentioned pointedly as some students made sub-optimal experiences with individual employees at the city and the university previously. Those employees were perceived as unfriendly or disinterested, lowering the impression of the institutions. At the city administration, negative experiences, e.g. at the foreigners' office, reduced the standing of the other city administration's offices as well. The decision making within the investigated services was described as a black box and not transparent. However, no special concerns regarding unfair or otherwise inappropriate decision making were stated.

Perceived Competency. While the goodwill was perceived as quite high, the perception of digital competency was described as significantly lower. While it was mentioned by international students that Germany as a country is "doing well for many years now" (Interview D) and therefore the German government probably does a good job, their specific outlook on digital competency was more critical. This was even more pronounced by the two German interviewees, who said that the recruitment of competent IT staff is crucial for the success of current and future digitalization projects. Both related this to the need for higher wages in the public sector to compete with private businesses which generally were perceived as more digitally competent. The digital competency of the university was generally perceived as higher. The corresponding interviewees explained this with the good reputation of the university in general and the positive experiences they made individually with the IT infrastructure. Furthermore, the awareness of the competency of the computer science/information systems lecturers as well as the university's IT department increased the perceived IT competency.

Perceived Data Security and Protection. Our interviewees in general were aware of the increased need for data protection if digital solutions are in place for public services. In general, the ability to provide data protection was judged as very high for various reasons. A general "faith" (Interview D) in the government to ensure security was mentioned, as well as the data protection policies and laws in place. Furthermore, the risk for the individual as "not someone who is very important" (Interview F) was perceived quite low. Next, a good digital solution was considered as even more secure than handing in forms and information to people in person but only if the solution would be developed in a competent and secure way. Otherwise, the potential benefits might be reversed and it would be even less secure. The need for a secure online authentication was mentioned connected with the concern of one international student that it "only works for German IDs" (Interview C) which would be disadvantageous for foreigners. One interviewee communicated a general skepticism regarding data security without being able to elaborate on the source of this skepticism. A general concern about using licensed software and cloud solutions without being able to control those (e.g., by a competent team developing open-source solutions for the government) was mentioned by one. Additionally, doubt was expressed that employees would be held accountable in case of mistakes, as it is difficult for them to get fired.

Information. In the end, all interviewees were able to find all information they needed to accomplish both processes. However, in particular regarding the enrolment at the university, not all the information was at one place and some instructions were not completely clear. Despite this, the interviewees were able to gather the information by themselves online or after contacting the university.

Communication. Contact to the city administration or the university was preferably done via an online channel. Only for short or urgent questions, e.g., if e-mails were unanswered, two interviewees utilized the phone to ask questions but otherwise all interviewees preferred to write an e-mail instead. This was due to various reasons: First, some prefer to not speak to a stranger if possible, one called it "being an introvert" (Interview B). This even hindered direct phone contact for Germans who otherwise would have no language barriers. Second, the language barrier and possible misunderstandings would intensify via the phone. Third, during in-person contact, especially with the city administration, the employees English skills were a barrier. Those were perceived as not sufficient by international students. Fourth, if the answer is not needed urgently the written answer was preferred as this might be more detailed (as employees might be in a hurry on the phone), as well as provable afterwards due to the documentation.

Convenience. The registration process at the city administration was overall described as sufficiently convenient. To go there in person was not perceived as a major pain point. Having the option to go to the city hall without an appointment was preferred, even if this corresponds with longer waiting times. Overall, as this process itself did not take too much time the interviewees had no further complaints. Regarding the enrolment at the university, the inconvenience of the process can be summed up into two major perceived problems: the processing time and the postal dispatch. The interviewees experienced different processing length but the waiting time between sending the required documents, the acceptance of those, and receiving the confirmation including payment information, was not perceived as convenient. This time was significantly increased if the letters were sent internationally. For some interviewees, the letters did not arrive in time or even got lost. Additionally, they needed to pay for insured international shipments. Overall, especially the international students who do not have the option to visit in person, would prefer to send all documents and receive all information digitally for an increased convenience of the enrolment.

Personal Preferences. During the research process we found that the interviewees have different personal preferences for their contact to administrations. Most interviewees prefer to do it online because of convenience, general communication fears or language barriers. However, the importance to at least have the option of personal contacts in today's digitalized world was highlighted. One interviewee would especially like to go to the administrations "in person just to have the feeling back" (Interview E) after the COVID-19 pandemic. Furthermore, the experiences of the international students in their home countries influenced their perception. For example, in Hungary additional processes can be done online e.g. a digital medical record, leading to a confusion why the German administration is less digital. Negative experiences of the Bulgarian student with her own government and the comparably positive image of the German government also increased the perceived competency of the German administration.

5 Discussion

For a qualitative approach, interpretation is needed. To analyze the **trusting beliefs and risk perceptions**, we focused on the different trust-related aspects instead of using the broad, general concept of trust as discussed prior (see Sect. 2). We also investigated these aspects for the different steps of service delivery, not focusing on the service process overall as done prior. In general, the citizens needed to do some steps offline and had to do the registration at the city in person and the enrolment at the university in person or via post. The international students had to do the enrolment via post as they were not yet on-site. Hence, the channel choice for the applications was forced. However, searching information and preparing applications could be done online for both services. A major identified risk perception was neither regarding the service provider itself, i.e. the local German authority, nor regarding the services, but related to the external source of international post as it was slow, unreliable, and expensive and therefore lowered the trust in the successful completion of the journey. Furthermore, all forms of personal communication, for example, due to information needs in the preparation phase, or otherwise during the application phase, were perceived as critical touchpoints to the administrations. In addition to personal preferences, concerns about language barriers lead to international students communicating as little as possible in person or via the phone. Comparable to other studies [1,3,4], we found the trust into the administration consisting of components like competency, integrity, and goodwill but also straightforward information and communication makes the interviewed students more comfortable to interact with the administrations and use their services.

The interviewees were asked whether they would like to do the services fully online. In regards to this, increased risks regarding data privacy and secure authentication methods were perceived. Although the local administrations' digital competency was not recognized as high by every interviewee, they believed in the integrity and the ability of the administrations to create a secure service once it was introduced. Here, the interview results indicate a relationship between trusting beliefs and risk perceptions as mirror images as described by Das and Teng [9]. International students expressed stronger competence beliefs in the German administation and subsequently also less risk perceptions than the German students. Further, reasons voiced were the strict regulations in data protection laws and the belief that the government administrations will comply with them. Additional concern decreasing effects were, that the administrations would receive the same information anyways only via another channel, or that the personal risks were considered low as one is not a person of interest. Overall, even though new concerns and perceived risks related to online services emerged, the level of trust in the public administration and the prospect of improved convenience would make all interviewees choose those online channels.

Considering the differences between the **city (local) and university (state) administration**, the former was generally perceived more positively, especially regarding providing good online services. The goodwill and integrity were mostly positive for both, with slight advances for the university. However, the more positive perception cannot be traced back to the different government level, as the interviewees did not highlight this. Two other major factors were found: First, the reputation of the university as a place of higher education, including diverse studies in the digital realm. Second, the higher number of touchpoints and prior experiences of students with the university compared to the city administration.

Overall, our interviews made similar discoveries for the **national and international students**. Same as Cabinakova et al. [6], we, too, found differences between our interviewees based on their background. However, the differences could mostly be traced back to specific individuals' experiences and language skills and were not necessarily cultural. Based on prior experiences, the German interviewees were more critical of the local administration's digital competency. Due to the services themselves, international students had more problems with less digitalized processes, as they cannot go there in person and delivery of documents is riskier and expensive. Furthermore, the language barrier was seen as a problem. Here, good digital solutions could help. Asynchronous communication like e-mails was mostly perceived as more comfortable than synchronous communication since it lowers the pressure of good German skills or employees' English skills. Communication problems did not directly lower the trust of international students but increased the uncertainty in dealing with the administrations.

Different **enablers** that foster the use of online service were found: the opportunity to reduce communication (not only but also in a foreign language), the increased convenience (less waiting time, no postal delivery), and a comprehensible overview of what steps are required by the citizens, including the alternatives to achieve them. These were also found to be the major downsides of the current service processes. However, **barriers** could be a missing awareness that services can be done online (as not all citizens are informed about their alternatives today), concerns about data protection, or unsecured or overly complicated authentication methods (not only but especially for foreigners).

The **citizen journey tool** [29] was found to be very effective for the service process analysis. It was especially useful during the introduction of the services to the interviewees and the visualization, as a foundation for the discussion of further service-relevant aspects (in this study - trust-related). The interviewees were able to keep the specific process in mind during their answers. Furthermore, it was easy to switch between the service as a complete process and the specific steps. Before and after using the citizen journey tool, no major revisions were necessary in order to utilize it for our research. In our opinion, this works as a proof of concept for the use of the citizen journey - on the one hand, to investigate general-purpose service experiences and on the other hand, as an analytical tool for more granular aspects, such as trust, in specific steps during service delivery.

In **future research** the enablers and barriers should be further investigated for a completely digital service. Also, the citizen journey tool could be used in additional research as an analytical tool for other services. Moreover, in-depth aspects other than trust, such as increased service automation or the role of digital service use on democracy, could be researched with it. Our study also suggests that the citizen journey tool could also be used as a visual aid that gives orientation to citizens on how to use a public service.

6 Conclusion

The research goal of this paper was to identify trust and risk perceptions in the citizen-administration-relationship in different phases of hybrid public services. We approached this by adopting the citizen journey tool as an analytical tool and conducting semi-structured interviews with national and international university students. The investigated hybrid public services are the residency registration at a German city administration and the enrolment at a public German university. Even though our study was limited to two services and six one-on-one interviews, we were able to deliver a proof of concept for the citizen journey to be used as an analytical tool. Thus, the citizen journey tool is suitable for the participation of citizens to obtain insights in how they perceive public service delivery. In conclusion, we have shown that trust and risk perceptions are relevant for the citizens' impressions of hybrid public services. Individual character traits and preferences and the personal background including previous experiences with German as well as home-country public administrations influence the perceptions of public service delivery. Furthermore, while having a positive perception about the general administrations' competency in general, our interviewees communicated perceived risks regarding the digital abilities of the administrations to provide secure digital services. They would still prefer to conduct them entirely online, due to existing convenience and communication issues, but also due to trusting beliefs in the goodwill and integrity of local German administrations.

References

1. Akkaya, C., Wolf, P., Krcmar, H.: The role of trust in E-government adoption: a literature review. In: Proceedings of the 16th Americas Conference on Information Systems, pp. 1–9. Lima, Peru (2010)
2. AlAwadhi, S., Morris, A.: Factors influencing the adoption of E-government services. J. Softw. **4**(6), 584–590 (2009)
3. Alsaghier, H., et al.: Conceptualising citizen's trust in e-government: application of Q methodology. Electrr. J. e-Govt. **7**(4), 295–310 (2009)
4. Alzahrani, L., Al-Karaghouli, W., Weerakkody, V.: Analysing the critical factors influencing trust in e-government adoption from citizens' perspective: a systematic review and a conceptual framework. Intern. Bus. Rev. **26**(1), 164–175 (2016)
5. Bologna Process. https://education.ec.europa.eu/levels/higher-education/inclusion-connectivity/bologna-process-european-higher-education-area. Accessed 16 Jan 2022
6. Cabinakova, J. et al.: The importance of culture, trust, and habitual patterns - determinants of cross-cultural e-government adoption. In: Proceedings of the 21st European Conference (2013)
7. Carter, L., Bélanger, F.: The utilization of e-government services: citizen trust, innovation and acceptance factors. Inf. Syst. J. **15**(1), 5–25 (2005)
8. Chopra, K., Wallace, W.A.: Trust in electronic environments. In: Proceedings of the 36th Hawaii International Conference on System Sciences (2003)
9. Das, T.K., Teng, B.S.: The risk-based view of trust: a conceptual framework. J. Bus. Psychol. **19**, 85–116 (2004)

10. Digital public services. https://futurium.ec.europa.eu/en/digital-compass/digital-public-services. Accessed 16 Jan 2022
11. Distel, B., et al.: The risk-based view of trust: a conceptual framework. In: Blöbaum, B. (ed.) Trust and Communication, pp. 163–184. Springer, Cham (2021). https://doi.org/10.1007/978-3-030-72945-5
12. Erasmus+. https://erasmus-plus.ec.europa.eu/.Accessed 16 Jan 2022
13. European Commission: eGovernment benchmark 2021: entering a new digital government era: background report. Publications Office of the European Union (2021)
14. European Commission: eGovernment benchmark 2021: entering a new digital government era: country factsheets. Publications Office of the European Union (2021)
15. Federal Ministry of the Interior and Community: Onlinezugangsgesetz (OZG). https://www.bmi.bund.de/DE/themen/moderne-verwaltung/verwaltungsmodernisierung/onlinezugangsgesetz/onlinezugangsgesetz-node.html. Accessed 11 Mar 2022
16. Federal Ministry of the Interior and Community: Residency registration. https://www.bmi.bund.de/EN/topics/administrative-reform/registration/registration-node.html. Accessed 16 Mar 2022
17. German Academic Exchange Service: Mobilität ausländischer Studierender, https://www.daad.de/de/der-daad/was-wir-tun/zahlen-und-fakten/mobilitaet-auslaendischer-studierender/. Accessed 9 Jan 2022
18. Graf-Drasch, V., Meindl, O., Voucko-Glockner, H.: Life is a journey in smart and sustainable districts. In: 17th International Conference on Wirtschaftsinformatik (WI 2022) Nuremberg, Germany (2022)
19. Grunow, D.: Die öffentliche Verwaltung in der modernen Gesellschaft. In: Verwaltung in Nordrhein-Westfalen, pp. 11–47. Aschendorff Verlag (2003)
20. Hofmann, S. et al.: Citizen-centered analysis of what public services are suitable for digital communication channels. In: Proceedings of Ongoing Research, Practitioners, Posters, Workshops, and Projects of the International Conference EGOV-CeDEM-ePart 2021, pp. 75–82 (2021)
21. Initiative D21 e.V.: eGovernment Monitor 2021: Staatliche Digitalangebote - Nutzung und Akzeptanz in Deutschland, Österreich und der Schweiz (2021)
22. Kaliontzoglou, A., et al.: A secure e-Government platform architecture for small to medium sized public organizations. Electron. Commer. Res. Appl. **4**(2), 174–186 (2005)
23. Lemon, K.N., Verhoef, P.C.: Understanding customer experience throughout the customer journey. J. Mark. **80**, 69–96 (2016)
24. Mayer, R.C., Davis, J.H., Shoorman, F.D.: An integrative model of organizational trust. Acad. Manag. Rev. **20**(3), 709–734 (1995)
25. Mensah, I.K., Zeng, G., Luo, C.: E-government services adoption: an extension of the unified model of electronic government adoption. SAGE Open **10**(2) (2020)
26. National Regulatory Control Council of the Federal Government of Germany: Monitor Digitale Verwaltung #6, https://www.normenkontrollrat.bund.de/nkr-de/aktuelles/monitor-digitale-verwaltung-6-1958280. Accessed 11 Mar 2022
27. Parent, M., Vandebeek, C.A., Gemino, A.C.: Building citizen trust through e-government. Gov. Inf. Q. **22**(4), 720–736 (2005)
28. Rana, N.P., Dwivedi, Y.K., Williams, M.D.: A meta-analysis of existing research on citizen adoption of e-government. Inform. Syst. Front. **17**(3), 547–563 (2015)
29. Scholta, H., Halsbenning, S., Distel, B., Becker, J.: Walking a mile in their shoes—a citizen journey to explore public service delivery from the citizen perspective. In: Viale Pereira, G., et al. (eds.) EGOV 2020. LNCS, vol. 12219, pp. 164–178. Springer, Cham (2020). https://doi.org/10.1007/978-3-030-57599-1_13

30. Sweeney, Arthur.D.P.: Electronic Government-Citizen Relationships. J. Inf. Technol. Polit. **4**(2), 101–116 (2007)
31. Teo, T.S.H., Srivastava, S.C., Jiang, L.: Trust and electronic government success: an empirical study. J. Mgmt. Inf. Syst. **25**(3), 99–131 (2008)
32. Wilson, J.: Essentials of Business Research: A Guide to Doing Your Research Project, 2nd edn. SAGE Publications (2014)
33. Wollmann, H.: German local government under the double impact of democratic and administrative reforms. In: Kersting, N., Vetter, A. (eds.) Reforming Local Government in Europe. Urban and Research International, vol. 4. VS Verlag für Sozialwissenschaften, Wiesbaden, pp. 85–112 (2003). https://doi.org/10.1007/978-3-663-11258-7_5
34. Yang, K.: Public administrators' trust in citizens: a missing link in citizen involvement efforts. Public Adm. Rev. **65**(3), 273–285 (2005)
35. Yera, A., et al.: Characterization of e-Government adoption in Europe. PLoS ONE **15**, 4 (2020)

ICT and Sustainability

Applications of Data-Driven Policymaking in the Local Energy Transition: A Multiple-case Study in the Netherlands

Devin Diran, Marissa Hoekstra(✉), and Anne Fleur van Veenstra

Netherlands Organization for Applied Scientific Research TNO, Anna Van Buerenplein 1, 2595 DA The Hague, The Netherlands

{devin.diran,marissa.hoekstra,annefleur.vanveenstra}@tno.nl

Abstract. The potential of the use of data to help improve policymaking is increasingly recognized by governments, especially to address societal challenges. One of those societal challenges is the energy transition, which happens for a large part at the local government level. However, within the literature little is known about which type of applications are utilized for data-driven policymaking. From government practice a plethora of data-driven applications are mentioned to be under development or in experimental phase, but not much is known on which applications are actually utilized by policymakers. Therefore, the aim of this exploratory study is to gain insight into how these data-driven applications support policymaking for the local energy transition. To investigate this, we perform a multiple-case study of four municipalities in the Netherlands. Using an analytical framework derived from an literature overview of data-driven applications for the local energy transition, we carry out four case studies of the local energy transition in the Netherlands. We find that they use data-driven applications throughout the whole policy cycle. However, a significant gap exists between data-driven applications to enable and accelerate the energy transition currently implemented, and the desired applications, but also the potential applications found in literature. We recommend future research pertaining to integrated and actionable adoption strategies in order to bridge this gap.

Keywords: Data-driven policymaking · Data-driven applications · Energy transition · Local government · Sustainable development goals · Policy cycle

Conference track: ICT and Sustainable Development Goals

1 Introduction

Governments increasingly recognize the potential of data for their policymaking process, since data-driven policy can contribute to enlarging transparency, legitimacy and efficiency of policy [1, 2]. In particular, government organizations aim to use data for policymaking processes that address societal challenges such as the energy transition,

Published by Springer Nature Switzerland AG 2022
R. Krimmer et al. (Eds.): ePart 2022, LNCS 13392, pp. 55–72, 2022.
https://doi.org/10.1007/978-3-031-23213-8_4

with the benefit to accelerate the development and implementation of these policies [3, 4]. From government practice a plethora of data-driven applications are mentioned to be in development or in experimental phase, and occasionally implemented. However, within the literature little is known on which applications are actually used by the policy-makers, and in particular how these data-driven applications are used to address energy transition policies.

Therefore, the aim of this exploratory study is to gain insight into how data-driven applications support policymaking in the local energy transition, via the following main research question: *Which data-driven applications are used to support policymaking in the local energy transition?*

In literature a variety of definitions for data-driven policymaking can be found, for instance [2, 5, 6], however a clear definition of data-driven policymaking is difficult to find. For this study we define data-driven applications for policymaking as *applications which generate, collect and utilize policy relevant data to identify patterns and trends towards feasible policy configuration, implementation and monitoring, and organizational efficiency in terms of time and resources needed to design and implement a policy solution.*

Within the literature several studies give more insight into the purpose of data-driven applications for the energy transition [7–9]. However, these studies often focus on the development of, or the utilization of an individual tool. A comprehensive overview of the potential of applications, and its connection to the policy cycle is lacking. In this study we are interested in the wider application of data-driven applications for the local energy transition. In order to answer the research question, this study conducts a multiple-case study of the local energy transition in the Netherlands, by looking at the four largest Dutch municipalities. The research design constitutes of two phases. First, via literature overview we explore the utilization of data-driven applications by governments for the energy transition on the local level, resulting in an analytical framework. Second, a multiple-case study is performed to derive the empirical insights within the Dutch context.

The paper is structured as follows. First, we provide an overview of the literature on data-driven applications that may support local energy policies. Subsequently, a description of the methodology used in this study is presented. Then, we present the results from a multiple-case study of the local energy transition in the Netherlands. The study concludes with a discussion of the findings, and with conclusions and recommendations for further research.

2 Literature Overview

2.1 Data-Driven Policymaking

The potential of the use of data to help improve policymaking, the operational process and service delivery towards citizens is increasingly recognized by local governments [1]. Data-driven policymaking builds on the notion of evidence-based policymaking [2, 5] and focusses on the use of data analytics, big data and open data for the three consecutive phases of policymaking, from the agenda setting phase to the policy implementation and evaluation phase [2, 5, 10]. Data-driven policy can contribute to enlarging transparency, legitimacy and efficiency of policy [2]. However, the adoption of data-driven policy differs per domain and some domains like the energy domain are behind with adopting data-driven applications [10–12]. Nevertheless, in light of the transition to a sustainable society, realizing trustworthy, affordable, sustainable and a just energy transition requires proper energy policies and strategies. Data on the environment, infrastructure, technology and economy and the behavior of citizens play an important role in developing and strengthening of the knowledge base of local government [2, 13].

There are a number of characteristics that can give insight into how data-driven policymaking by local governments is utilized, such as the purpose of an application, the type of data analytics used and the involvement of stakeholders. The next paragraphs will provide more insight into these characteristics.

2.2 Purposes of Applications that Support the Policymaking Cycle

Local governments play a key role in implementing policies that support the transition to a sustainable society [14–16]. These policies are often focused on reducing emissions by reducing energy consumption or by switching to the use of alternative renewable energy resources [14]. The objective of this section is to provide a synthesis of the purposes of applications found in literature that support the policymaking cycle. The results are presented in Table 1.

Table 1. Synthesis of the purposes of applications found in literature that support the policymaking cycle.

Phase in policy cycle	Purpose of application	Source
Agenda setting & problem definition	Assessing the transition challenge via sustainable energy potential mapping and energy demand mapping	[7, 17, 18]
Policy design and demonstration	Building renovation planning Heating and cooling infrastructure planning Citizen and stakeholder engagement Energy transition strategy and policy selection and planning	[8, 19, 20] [21, 22] [23–25] [26, 27]
Policy implementation, monitoring and evaluation	Implementation monitoring on the achieved results Policy evaluation	[14, 15] [28]

For example, data-driven applications can help local policy makers throughout the policy cycle with agenda setting, identifying the problem definition, policy design and policymaking, policy implementation and evaluation [2, 13]. Another specific purpose is the monitoring of policies, as monitoring of local energy policies can help give insight into the progress that is being made with regard to these policies [14]. In addition, data-driven applications can be used for identifying the potential of renewable energy

[17, 29, 30]; applications for mapping the energy demand [8, 19, 20, 26]; applications for planning renovations of buildings [7, 31]; applications for the planning of heat and cooling networks [24, 25] and applications focused at social innovation and citizen and stakeholder engagement [23, 25, 27, 32, 33].

Over these applications catering to the various phases of the policymaking cycle, a variety of tools can be identified with an enabling role for the application to achieve its purpose. For example, Matheus et al. [34] conducted a study into how data-driven dashboards can contribute to decision-making in smart cities. Dashboards can help governments to interact with citizens and to empower these citizens. They argue that a dashboard should not only be developed to provide information, but also to gather feedback from citizens and to stimulate interaction between citizens and government [34]. Another type of tool that can be used by policy makers are energy models. An energy model is "*a computer model of an energy system that introduces a structured way of thinking about the implications of changing parts of the system*" [35]. Energy models can be used for designing energy transition policy, as it may help to understand the complexity of the energy sector [35].

Mapar et al. [15] conducted a study that looks beyond the energy domain, and examined which indicators and data should be adopted to evaluate health, safety and environmental operations in municipalities in megacities. The study shows that by combining and linking several dimensions such as health, safety and environmental dimensions through an integrated organizational performance tool, more insight can be given into the sustainability performance of a city as a whole [15]. There currently is a gap in the literature on Digital Governance in the availability of the state of the art data-driven applications, models and tools and the adoption of these applications in local energy policymaking [10].

2.3 Type of Data Analytics

Moreover, the type of data analytics used can provide insight into the type of support catered to policymaking. Gartner's Data Analysis Value model identifies four categories of data analysis that provide value for policy and decision-making, with increasing complexity and value with each stage [36]: 1) *Descriptive analysis*: what is the current situation and what has occurred?, 2) *Diagnostic analysis*: What is causing this?, 3) *Predictive analysis*: What will happen in the future?, and 4) *Prescriptive analysis*: How can change/desired outcome be accomplished? In addition to the type of support needed by policymaking, this model can give an indication of the maturity of data-driven applications, as descriptive analysis is a basic form of data analysis, whereas the use of prescriptive analysis is more complex and requires more advanced skills and methods.

2.4 Stakeholder Engagement

Furthermore, the development of data-driven applications often occurs within a data ecosystem that consists of multiple users and stakeholders that manage different data sources [4]. Furthermore, partnerships between policymakers and stakeholders of the government offer various benefits like coalition building, exchange of knowledge and gaining access to resources such as data and capacity [37]. In particular, the energy transition requires support from various stakeholders in society, since the energy transition is a long and gradual process that affects many parties and stakeholders [38].

3 Case Study Methodology

In this study a multiple-case study is deployed to derive the empirical insights within the Dutch context. Four cases are subject of data collection and analysis, these cases are the four largest municipalities in the Netherlands: the municipality of Amsterdam, The Hague, Rotterdam and Utrecht. The selection criteria resulting in these four cases address the maturity of local energy transition policy and data-driven applications within policymaking and public service provision. All four municipalities are frontrunners in local energy transition policy and data-driven applications within policymaking and public service provision.

Interviews were the main method used for data gathering, in combination with desk research of background information, e.g. policy documents, provided by the case cities. The semi-structured interview protocol was established by taking into account the information gathered in the literature review. Between June 2021 and September 2022 a total of nine online interviews of approximately one hour took place. For each municipality a policy officer or program manager and a data scientist or policy researcher were interviewed (see Table 2). The goal of these interviews were to gain more insight into the current and desired role of data in the local energy transition, how this is realised in practice and which applications are used by municipalities.

Table 2. Interview respondents

Municipality	Respondent role
The Hague	Policy Advisor Energy Transition
The Hague	Policy Researcher
The Hague	Project leader public services energy transition and sustainability
Amsterdam	Coordinator Data and Monitoring Sustainability
Amsterdam	Policy Researcher
Amsterdam	Account manager Research, Information & Statistics
Rotterdam	Process manager and advisor sustainability
Rotterdam	Program manager Digital City Rotterdam
Utrecht	Policy Advisor area based sustainability
Utrecht	Data Scientist Energy Transition

To examine the results, the following procedures were followed. First, an interview protocol was created, and the interviews were then carried out in accordance with the interview protocol. Following the interviews, a report on the interviews was created and validated with the respondents.

The data was then evaluated after coding of the applications, opportunities, challenges, added value, and preconditions, using QDA miner software. The data analysis framework encompasses data-driven application which consist of three elements:

- **The purpose of the data-driven application** i.e. to which policies is the data application contributing and with which objective.
- **The type of data analytics** utilized in the data-based application, namely: 1) *Descriptive analysis*, 2) *Diagnostic analysis*, 3) *Predictive analysis*, and 4) *Prescriptive analysis*.
- **The user-group/audience** for which the application is developed, namely: policy makers, city stakeholders (businesses or civil society) or citizens.

It should be noted that, in line with the knowledge gap, the applications included and studied in this paper are applications which are actually adopted in the policymaking cycle to support energy transition policy. The next chapters present the finding of this research.

4 Results Case Studies

Where the previous section addresses the literature on data-driven applications for the energy transition, this section sheds light on the results of the multiple-case study encompassing the four largest municipalities in the Netherlands: Amsterdam, The Hague, Utrecht and Rotterdam.

4.1 Background: Local Energy Transition in the Netherlands

Local governments play a key role in implementing policies that support the transition to a sustainable society [14–16]. In the Netherlands, municipalities are obliged to establish a Transition Vision Heat (TVH) encompassing the disconnection of cities from natural gas towards sustainable sources of heating, a Neighborhood Implementation Plan (NIP) further detailing the transition in the neighborhood, and contribute towards Regional Energy Transition Strategies (RETS) which combine the municipal commitments in local renewable energy generation on the regional level [39]. Table 3 summarizes the policy needs of local governments, and how this translates towards the need for data-driven support.

Table 3. Policy and public service needs for data-driven support in the Netherlands. Adapted from [39, 40]

Policy or public service need	Main activities
Phase 1: Strategy development, entailing the sequence of districts to disconnect from natural gas by 2030 and the preferred alternative for sustainable heating (Transition Vision Heat)	- Exploration of the current situation, e.g. with GIS tools and dashboards - Techno-economic analysis of heating alternatives to decide on the heating solution, with assessment models
Phase 2: District Implementation Plans	- Developing district implementation plans. This calls for detailed data on the heterogenous building stock and citizens - Implement, monitor and adapt plans
Phase 3: Contribute in Regional Strategy for Sustainable Electricity Generation	- Determine municipal potential to generate renewable electricity - Negotiate commitment to regional renewable electricity supply and what the municipality needs to accomplish this

The establishment and implementation of these policies and public services bring various knowledge needs for policymakers pertaining to the economic, social, and environmental aspects of the sustainability transitions [4, 14, 41]. To close these knowledge gaps the G4 develop and utilize data-driven applications. In the coming paragraphs, the data-driven applications in the four case studies are presented after analysis on three variables, namely the purpose of the application, type of analytics utilized in the application, and the user-group for which the application is developed.

4.2 Amsterdam

The city of Amsterdam (population: 1.158.000, area: 219 km^2) is widely known as a frontrunner on IT adoption within the municipal organization for policymaking and the provision of public services [42]. As such, the municipality houses a data lab, which is supported by the Department of Research, Information, and Statistics (RIS) of around 30 FtE. The department of RIS, together with the Chief Innovation Office are working on the establishment of expertise and capacity for data applications, privacy and information security. In addition, Amsterdam is a leader among Dutch municipalities and in Europe regarding sustainability policy development and implementation. The sustainability department, is concerned with the city's energy transition and has a data and monitoring coordinator to encourage data-driven work in the energy transition.

Table 5 presents the applications found in Amsterdam addressing the roadmap towards a climate neutral city, circularity, but also the impact on employment and actions required from the local businesses.

The applications of data and digital technologies predominantly utilize descriptive and diagnostic data analytics, while predictive analytics are rare. All applications are developed for the local policy makers, while specific applications (Roadmap Climate

Neutral Amsterdam, Climate Report Amsterdam, Program Sustainable Businesses and Analysis Employment Energy Transition) are also developed towards city stakeholders and citizens as they require collective action and support.

About the added value of working with data, a respondent from the municipality states the following:

"The energy transition is a theme with different topics, in which utilizing a lot of information is necessary to make good decisions. We are working on complex questions. By connecting various perspectives you try to find an answer: interests and support of stakeholders and citizens, how certain buildings are build, how does the energy system look like. This makes it so complex that you can only develop and build it with data." (respondent municipality of Amsterdam, translated from Dutch to English) (Table 4).

Table 4. Data-driven applications supporting policymaking in Amsterdam

Application	Purpose of application	Type of analytics	Target audience/user
Analytics for the Routekaart Klimaatneutraal Amsterdam (Roadmap Climate Neutral Amsterdam)	Transition challenge assessment via sustainable energy potential and energy demand mapping; Citizen and Stakeholder Participation; System interdependency Analysis; Energy transition strategy and policy selection and planning	Descriptive; Diagnostic; Predictive	Policy Makers; City Stakeholders (Businesses and Civil Society Organizations); Citizens
Analytics for the Klimaatrapport Amsterdam (Climate Report Amsterdam)	Policy Implementation, Monitoring and Evaluation	Descriptive	Policy Makers; City Stakeholders; Citizens
Analytics to support the Programma Verduurzaming Zakelijke Markt (Program Sustainable Businesses)	Transition challenge assessment via sustainable energy potential and energy demand mapping; Citizen and Stakeholder Participation; Energy transition strategy and policy selection and planning	Descriptive	Policy Makers; City Stakeholders

(*continued*)

Table 4. (*continued*)

Application	Purpose of application	Type of analytics	Target audience/user
Analytics Employment Energy Transition	Transition challenge assessment via sustainable energy potential and energy demand mapping; Energy transition strategy and policy selection and planning	Descriptive; Diagnostic	Policy Makers
Data for Circulaire Monitor (Monitor Circularity)	Policy Implementation, Monitoring and Evaluation	Descriptive	Policy Makers

4.3 Rotterdam

Rotterdam (population: 652.000, area: 324 km^2) is researching the potential of digitalization for the future city through the Digital City initiative. The construction of a digital Open Urban Platform with a three-dimensional Digital Twin of the city is at the heart of this effort. Knowledge is currently being gathered through projects and pilots, which will be used to further accelerate these improvements. Rotterdam is regarded as pioneer for Digital Twins and digital innovation in the public sector, recognized with a Global UNESCO award in 2021 [43].

Table 6 presents the applications found in Rotterdam, with the main applications of the WHAT map (which option for sustainable heat is promising per district?) and the WHEN map (when will it be the turn of which district?) constituting Rotterdam's Transition Vision Heat (TVH). In addition, a Digital Twin Sustainability and an Urban Data Platform are established with important enabling roles towards the utilization of data and digital technologies to support the energy transition decision making. About this Digital Twin, a respondent from the municipality of Rotterdam states the following:

"The Digital Twin will be used by the municipality of Rotterdam as a basis for the development of new applications, to test how data and digital technologies can be used to engage stakeholders, in particular citizens, with regards to the sustainability transition." (respondent municipality of Rotterdam, translated from Dutch to English).

The applications of data and digital technologies utilize descriptive, diagnostic and prescriptive data analytics.

All applications are widely developed for local policy makers, while also targeting the city stakeholders and citizens. City stakeholders and citizens have an important role towards the enabling applications in the Urban Data Platform and Digital Twin to acquire and share the necessary data. Moreover, data applications for the WHAT and WHEN Map include citizens and city stakeholders to ensure the necessary support for these policy choices.

Table 5. Data-driven applications supporting policymaking in Rotterdam

Application	Purpose of application	Type of analytics	Target audience/user
Analytics for the WAT Kaart (What Map)	Transition challenge assessment via sustainable energy potential and energy demand mapping; Energy transition strategy and policy selection and planning	Descriptive; Diagnostic; Prescriptive	Policy Makers; Citizens; City Stakeholders
Analytics for the WANNEER Kaart (When Map)	Energy transition strategy and policy selection and planning	Descriptive; Diagnostic; Prescriptive	Policy Makers; Citizens; City Stakeholders
Urban Data Platform	Enabling: Data sharing, Data storage, Data collection	N/A	Policy Makers; Citizens; City Stakeholders

4.4 The Hague

In 2021 the municipality of The Hague (population: 550.000, area: 98 km^2) founded the energy transition data lab, to collaborate with parties interested in the energy transition. The Municipality of The Hague's energy transition program team focuses on, among other things, the city's energy transformation.

Table 6 presents eight types of data-driven applications found in the municipality of The Hague. The type of data analytics used in each application varies from descriptive analytics to predictive analytics in some cases. In addition the majority of applications are used by policy officers, although in some case stakeholders and citizens are the intended user as well (e.g. data lab energy transition). Moreover, the types of applications vary from supporting activities throughout the whole policy cycle to stimulating citizen and stakeholder participation. During the interviews, the municipality of The Hague, Utrecht and Amsterdam mention the need for access to social data. A respondent from the municipality of The Hague states the following on social data:

"*You need to closely look at a neighborhood, through the social data. What is the willingness to invest? What issues are going on in this neighborhood? I am looking for social data, to gain insight into social effects and to take those social effects into account for your policy decisions.*" (Respondent from the municipality of The Hague, translated from Dutch to English).

Table 6. Data-driven applications supporting policymaking in The Hague

Application	Purpose of application	Type of analytics	Target audience/users
Analytics in the Data Lab Energy Transition	Transition challenge assessment via sustainable energy potential and energy demand mapping; Citizen and Stakeholder Participation; Energy transition strategy and policy selection and planning	Descriptive; Diagnostic	Policy makers; Citizens; City Stakeholders
CO2 Monitor	Energy transition strategy and policy selection and planning; Citizen and Stakeholder Participation; Policy Implementation, Monitoring and Evaluation	Descriptive; Diagnostic	Policy makers
Analytics to assess and predicting the efficacy of policies, e.g. subsidies for solar energy and insulation	Energy transition strategy and policy selection and planning; Policy Implementation, Monitoring and Evaluation	Predictive	Policy makers
Analytics for tailored policy	Citizen and Stakeholder Participation	Descriptive	Policy makers; citizens
Analytics for Neighborhood scan: Baseline measurement of a neighborhood	Transition challenge assessment via sustainable energy potential and energy demand mapping; Citizen and Stakeholder Participation; Energy transition strategy and policy selection and planning;	Descriptive	Policy makers
Analytics for Dakenjacht (sustainable energy on roofs)	Transition challenge assessment via sustainable energy potential and energy demand mapping; Citizen and Stakeholder Participation	Descriptive; Diagnostic	Policy makers; City stakeholders

(*continued*)

Table 6. (*continued*)

Application	Purpose of application	Type of analytics	Target audience/users
Data for transition dashboard	Policy Implementation, Monitoring and Evaluation	Descriptive	Policy makers

4.5 Utrecht

The city of Utrecht (population: 360.000, area: 99 km^2) is famed for its in-house data brigade. Data scientists form an integral part of the energy transition team. Furthermore, the municipality of Utrecht has created an open data platform focusing on a variety of topics including culture, economy, finance, housing, agriculture, migration, nature and the environment, physical space and infrastructure, and mobility.

Table 7 presents seven types of data-driven applications found in the municipality of Utrecht. The majority of applications have forms of descriptive analytics and are used by policy officers of the municipality. In addition, many applications are aimed at energy transition strategy and policy selection and planning, mapping the potential for sustainable energy resources in the city, citizen and stakeholder participation, monitoring of policy and goals. A new application is aimed at system interdependency analysis, in which dependencies of the energy transition are mapped. A respondent from the municipality of Utrecht states the following on this:

"By gaining insight into these interdependencies, you gain insight into the consequences of a decision. The municipality is constantly working on this, to take this into account in the decision-making process. For example, if the politicians oppose biomass, I show the scheme of dependencies to illustrate that without biomass there are not many other options and that e.g. hundred wind mills need to be build. This results in considerations and discussions that are part of the decision-making process." (Respondent municipality Utrecht, translated from Dutch to English).

Table 7. Data-driven applications supporting policymaking in Utrecht

Application	Purpose of application	Type of analysis	Target audience/users
Scenario analytics for Transition Vision Heat (TVH)	Energy transition strategy and policy selection and planning;	Descriptive; Diagnostic; Predictive	Policy makers
Analytics for the sustainable resources scan TVH	Transition challenge assessment via sustainable energy potential mapping;	Descriptive	Policy makers

(*continued*)

Table 7. (*continued*)

Application	Purpose of application	Type of analysis	Target audience/users
Participative Value Evaluation analytics for participation in TVH decision making	Citizen and Stakeholder participation; Energy transition strategy and policy selection and planning;	Descriptive; Diagnostic	Policy makers
Data for participation level measurement	Citizen and Stakeholder Participation	Descriptive	Policy makers
Dashboard utilization at citizen workshops	Citizen and Stakeholder participation; System interdependency Analysis	Descriptive; Diagnostic	Policy makers; Citizens
Data to monitoring the energy transition e.g. solar panels on roofs, connections to gas and district heating	Energy transition strategy and policy selection and planning; Policy Implementation, Monitoring and Evaluation	Descriptive; Diagnostic	Policy makers

5 Findings and Discussion

Table 8 presents an overview the identified purposes for which data-driven applications can be used for local energy policy. In addition, some purposes were identified in the case studies with little to no attention in the literature, namely: system interdependency analysis and applications that enable data sharing, data storage and data collection within the municipality.

The cases illustrate that the four largest Dutch municipalities utilize data-based applications throughout the policy cycle. These applications largely coincide with the applications found in literature, e.g.: transition challenge assessment via sustainable energy potential and energy demand mapping [7, 17, 18], energy transition strategy and policy selection and planning [26, 27] pertaining to e.g. heat networks and other infrastructure [21, 22], Social innovation for citizen and stakeholder engagement [23–25], and policy monitoring and evaluation analysis and reporting [14, 28]. Especially policy monitoring is a popular purpose of application for all municipalities.

Table 8. Overview of case results purpose of data-driven applications

Phase in policy cycle	Purpose of application	Amsterdam	Rotterdam	The Hague	Utrecht
Agenda setting & problem definition	Assessing the transition challenge via sustainable energy potential - and energy demand mapping	X	X	X	X
Policy design and policy making	Energy transition strategy and policy selection and planning	X	X	X	X
	Social innovation for citizen and stakeholder engagement	X	X	X	X
	System Interdependency analysis	X	X	X	X
Policy implementation, monitoring and evaluation	Policy implementation monitoring in dashboards and reports	X	X	X	X
	Policy evaluation analysis	X		X	
	Enabling: Data sharing, Data storage, Data collection		X		X

Table 9. Overview of case results type of data analytics

Type of data analytics	Amsterdam	Rotterdam	The Hague	Utrecht
Descriptive analytics	5/5	2/3	6/7	6/6
Diagnostic analytics	2/5	2/3	3/7	4/6
Predictive analytics	1/5		1/7	1/6
Prescriptive analytics	0/1	2/3		

Table 9 presents an overview of the type of data analytics identified. The majority of applications use basic descriptive data analytics, indicating that there are still many steps to take by municipalities before they can utilize the full potential of data-driven applications. Diagnostic and predictive analytics are also emerging, although to a lesser extent. Municipalities do share the need for predictive analytics in particular to assess the impact of policies and due to the long term nature of the energy transition with many uncertainties and knowledge gaps towards 2050.

Table 10. Overview of case results type of users

Type of users	Amsterdam	Rotterdam	The Hague	Utrecht
Policy makers	5/5	3/3	7/7	6/6
City stakeholders	3/5	3/3	2/7	0/6
Citizens	1/5	3/3	2/7	1/6

Table 10 provides an overview of the type of users. End-users of these applications are primarily policy makers. However, there also a number of applications that target both policy makers and city stakeholders. In most of these applications policy makers and city stakeholders collaborate with the purpose to share knowledge, data and resources, confirming the importance and benefits of collaborations between policymakers and its stakeholders [37]. Nevertheless, only few applications involve citizen engagement, or social data on citizens, even though Utrecht, The Hague and Amsterdam do indicate the need for social data.

This study encompasses two limitations. First, we focus on four large municipalities in the same country. Knowledge gaps remain if these applications are representative for smaller municipalities, and approached in other countries. Second, this study solely focusses on applications in the local energy policy domain, sustainability and the energy transition are much broader, therefore we suggest that future research takes into account other aspects of sustainability such as health and safety [15]. We recommend structured literature review and more case-studies to better underpin the research gaps identified in this exploratory study, among others by including smaller cities, the national and provincial level of government and a broader geographical coverage.

6 Conclusion

Local governments have a crucial role in the operationalization and realization of the energy transition. In order to gain insight into how data-driven applications support policymaking for the local energy transition, this study conducted a multiple-case study of the local energy transition in the Netherlands. It aimed to answer the question: *Which data-driven applications are used to support policymaking in the local energy transition?* We find that the four largest municipalities in the Netherlands: Amsterdam, Rotterdam, The Hague and Utrecht use data-driven applications throughout the entire policy cycle and for a variety of purposes, with the most common applications targeting the following purpose: transition challenge assessment via the mapping of energy demand and sustainable energy potential; citizen and stakeholder participation; energy transition strategy and policy selection and planning and policy monitoring and evaluation. The majority of applications use basic descriptive data analytics, while diagnostic and predictive analytics are emerging. End-users of these applications are primarily policy makers and few applications target citizens and other stakeholders of the municipality. All in all, a significant gap exists between data-driven applications to enable and accelerate the energy transition currently implemented, and the desired applications, but also the potential applications found in literature. We recommend future research pertaining to integrated and actionable adoption strategies in order to bridge this gap.

References

1. Ali, H., Titah, R.: Is big data used by cities? Understanding the nature and antecedents of big data use by municipalities. Gov. Inf. Q. **38**(4) (2021). https://doi.org/10.1016/j.giq.2021.101600
2. Veenstra, A.F., Kotterink, B.: Data-driven policy making: the policy lab approach. In: Parycek, P., et al. (eds.) ePart 2017. LNCS, vol. 10429, pp. 100–111. Springer, Cham (2017). https://doi.org/10.1007/978-3-319-64322-9_9
3. Manfren, M., Nastasi, B., Groppi, D., Garcia, D.A.: Open data and energy analytics - an analysis of essential information for energy system planning, design and operation. Energy **213**, 118803 (2020). https://doi.org/10.1016/j.energy.2020.118803
4. Diran, D., Hoppe, T., Ubacht, J., Slob, A., Blok, K.: A Data ecosystem for data-driven thermal energy transition: reflection on current practice and suggestions for re-design. Energies **13**(2), 444 (2020)
5. Giest, S.: Big data for policymaking: fad or fasttrack? Policy Sci. **50**(3), 367–382 (2017). https://doi.org/10.1007/s11077-017-9293-1
6. Höchtl, J., Parycek, P., Schöllhammer, R.: Big data in the policy cycle: policy decision making in the digital era. J. Organ. Comput. Electron. Commer. **26**(1–2), 147–169 (2016)
7. Gupta, R., Gregg, M.: Local energy mapping using publicly available data for urban energy retrofit. In: Dastbaz, M., Gorse, C., Moncaster, A. (eds.) Building Information Modelling, Building Performance, Design and Smart Construction, pp. 207–219. Springer, Cham (2017). https://doi.org/10.1007/978-3-319-50346-2_15
8. Chen, Y., Hong, T., Piette, M.A.: Automatic generation and simulation of urban building energy models based on city datasets for city-scale building retrofit analysis. Appl. Energy **205**, 323–335 (2017)
9. Henrich, B.: The Use of Energy Models in Heating Transition Decision Making: Insights from Ten Heating Transition Case Studies in the Netherlands. Delft University of Technology, Delft (2020)
10. Diran, D., Veenstra, A.F.: Towards data-driven policymaking for the urban heat transition in The Netherlands: barriers to the collection and use of data. In: Pereira, G.V., et al. (eds.) EGOV 2020. LNCS, vol. 12219, pp. 361–373. Springer, Cham (2020). https://doi.org/10.1007/978-3-030-57599-1_27
11. Poel, M., Meyer, E.T., Schroeder, R.: Big data for policymaking: great expectations, but with limited progress? Policy Internet **10**(3) (2018). https://doi.org/10.1002/poi3.176
12. Pfenninger, S., DeCarolis, J., Hirth, L., Quoilin, S., Staffell, I.: The importance of open data and software: is energy research lagging behind? Energy Policy **101**, 211–215 (2017)
13. European Commission: Quality of public administration: a toolbox for practitioners. Publications Office of the European Union (2017). https://doi.org/10.2767/879305
14. Kaselofsky, J., März, S., Schüle, R.: Bottom-up monitoring of municipal energy and climate policy: more than an alternative to top-down approaches? Prog. Ind. Ecol. **8**(4) (2014). https://doi.org/10.1504/PIE.2014.066804
15. Mapar, M., Jafari, M.J., Mansouri, N., Arjmandi, R., Azizinejad, R., Ramos, T.B.: Sustainability indicators for municipalities of megacities: integrating health, safety and environmental performance. Ecol. Ind. **83** (2017). https://doi.org/10.1016/j.ecolind.2017.08.012
16. Soares, D., Sarantis, D., Lameiras, M.: Improve cities resilience and sustainability through e-government assessment (2018)

17. Fremouw, M., Bagaini, A., De Pascali, P.: Energy potential mapping: open data in support of urban transition planning. Energies **13**(5), 1264 (2020)
18. Ramachandra, T.V., Shruthi, B.V.: Spatial mapping of renewable energy potential. Renew. Sustain. Energy Rev. **11**(7), 1460–1480 (2007)
19. Linder, L., Vionnet, D., Bacher, J.P., Hennebert, J.: Big building data-a big data platform for smart buildings. Energy Procedia **122** (2017). https://doi.org/10.1016/j.egypro.2017.07.354
20. Mathew, P.A., Dunn, L.N., Sohn, M.D., Mercado, A., Custudio, C., Walter, T.: Big-data for building energy performance: lessons from assembling a very large national database of building energy use. Appl. Energy **140**, 85–93 (2015)
21. Dalipi, F., Yayilgan, S.Y., Gebremedhin, A.: A cloud computing framework for smarter district heating systems. In: 2015 IEEE 12th International Conference on Ubiquitous Intelligence and Computing and 2015 IEEE 12th International Conference on Autonomic and Trusted Computing and 2015 IEEE 15th International Conference on Scalable Computing and Communications and Its Associated Workshops (UIC-ATC-ScalCom), pp. 1413–1416 (2015)
22. Noussan, M., Jarre, M., Poggio, A.: Real operation data analysis on district heating load patterns. Energy **129**, 70–78 (2017)
23. Kontokosta, C.E., Reina, V.J., Bonczak, B.: Energy cost burdens for low-income and minority households: evidence from energy benchmarking and audit data in five US cities. J. Am. Plan. Assoc. **86**(1), 89–105 (2020)
24. Li, F.G.N., Strachan, N.: Take me to your leader: using socio-technical energy transitions (STET) modelling to explore the role of actors in decarbonisation pathways. Energy Res. Soc. Sci. **51**, 67–81 (2019). https://doi.org/10.1016/j.erss.2018.12.010
25. Peterson, M., Feldman, D.: Citizen preferences for possible energy policies at the national and state levels. Energy Policy **121**, 80–91 (2018)
26. van den Dobbelsteen, A., Roggema, R., Tillie, N., Broersma, S., Fremouw, M., Martin, C.L.: Urban energy masterplanning—approaches, strategies, and methods for the energy transition in cities. In: Urban Energy Transition, pp. 635–660. Elsevier (2018)
27. Al-Lawati, A., Barbosa, L.: A framework for intelligent policy decision making based on a government data hub. In: Alexandrov, D.A., Boukhanovsky, A.V., Chugunov, A.V., Kabanov, Y., Koltsova, O., Musabirov, I. (eds.) DTGS 2019. CCIS, vol. 1038, pp. 92–106. Springer, Cham (2019). https://doi.org/10.1007/978-3-030-37858-5_8
28. Wang, D.Y.C., Trappey, A.J.C., Trappey, C.V., Li, S.J., et al.: Intelligent and concurrent analytic platform for renewable energy policy assessment using open data resources. In: Moving Integrated Product Development to Service Clouds in the Global Economy, pp. 781–789 (2014)
29. Kramers, L., Van Wees, J.-D., Pluymaekers, M.P.D., Kronimus, A., Boxem, T.: Direct heat resource assessment and subsurface information systems for geothermal aquifers; the Dutch perspective. Netherlands J. Geosci. **91**(4), 637–649 (2012)
30. Schiel, K., Baume, O., Caruso, G., Leopold, U.: GIS-based modelling of shallow geothermal energy potential for CO_2 emission mitigation in urban areas. Renew. Energy **86**, 1023–1036 (2016)
31. Miller, C.: Predicting success of energy savings interventions and industry type using smart meter and retrofit data from thousands of non-residential buildings. In: Proceedings of the 4th ACM International Conference on Systems for Energy-Efficient Built Environments, p. 17 (2017)
32. Truong, N.B., Cao, Q.H., Um, T.-W., Lee, G.M.: Leverage a trust service platform for data usage control in smart city. In: 2016 IEEE Global Communications Conference (GLOBECOM), pp. 1–7 (2016)

33. Li, R., Crowe, J., Leifer, D., Zou, L., Schoof, J.: Beyond big data: social media challenges and opportunities for understanding social perception of energy. Energy Res. Soc. Sci. **56**, 101217 (2019)
34. Matheus, R., Janssen, M., Maheshwari, D.: Data science empowering the public: data-driven dashboards for transparent and accountable decision-making in smart cities. Gov. Inf. Q. **37**(3) (2020). https://doi.org/10.1016/j.giq.2018.01.006
35. Henrich, B., Hoppe, T., Diran, D., Lukszo, Z.: The use of energy models in local heating transition decision making: insights from ten municipalities in The Netherlands. Energies **14**(2), 423 (2021)
36. Schlegel, K., Sallam, R.L., Yuen, D., Tapadinhas, J.: Magic Quadrant for Business Intelligence and Analytics Platforms. Gartner (2013)
37. Eden, L., Wagstaff, M.F.: Evidence-based policymaking and the wicked problem of SDG 5 Gender Equality. J. Int. Bus. Policy **4**(1), 28–57 (2020). https://doi.org/10.1057/s42214-020-00054-w
38. Berger, L., Bréchet, T., Pestiaux, J., van Steenberghe, V.: Case-study - the transition of Belgium towards a low carbon society: a macroeconomic analysis fed by a participative approach. Energy Strateg. Rev. **29** (2020). https://doi.org/10.1016/j.esr.2020.100463
39. Diran, D., Henrich, B., Geerdink, T.: Supporting municipal energy transition decision-making (2020)
40. Diran, D., van Veenstra, A.F., Brus, C., Geerdink, T.: Data voor de Transitievisie Warmte en Wijkuitvoeringsplannen. Den Haag (2020)
41. Vringer, K., de Vries, R., Visser, H.: Measuring governing capacity for the energy transition of Dutch municipalities. Energy Policy **149** (2021). https://doi.org/10.1016/j.enpol.2020.112002
42. Noori, N., Hoppe, T., de Jong, M.: Classifying pathways for smart city development: comparing design, governance and implementation in Amsterdam, Barcelona, Dubai, and Abu Dhabi. Sustainability **12**(10), 4030 (2020)
43. TheMAYOR.eu: UNESCO recognises Rotterdam as a digital pioneer (2021). https://www.themayor.eu/en/a/view/unesco-recognises-rotterdam-as-a-digital-pioneer-7702

The Great Divide: Empirical Evidence of a Decoupling of Digital Transformation and Sustainability

Livia Norström(✉), Johan Magnusson, and Vasili Mankevich

University of Gothenburg, Forskningsgången 6, 417 56 Gothenburg, Sweden
livia.norstrom@ait.gu.se

Abstract. Digital transformation is increasingly prioritized within the public sector to assure sustained relevance. At the same time, sustainable development goals (SDGs) are increasingly addressed in the strategies and missions of public sector actors. Previous research highlights that sustainability requires integration into internal procedures and governance structures (e.g., accounting) to avoid running the risk of merely being ceremonial and green washing. With digital transformation deemed critical for the public sector, we would hence expect to see an integration of SDGs in digital initiatives. This study answers the research question of how SDGs are integrated into digital transformation initiatives. We answer the question through purposive sampling of digital initiatives within a large, Swedish municipality, where each initiative is categorized on the notion of which SDGs they are related to. The findings show that there is a decoupling of sustainability in digital transformation initiatives, that risks leading to directly detrimental effects for both the digital transformation of the public sector as well as for sustainability. This is discussed through integration with previous literature with the intent of identifying future avenues for research and recommendations for practice.

Keywords: Digital transformation · Sustainable development goals · SDG · Responsible information systems · Digital government

1 Introduction

The public sector is experiencing a shift toward digital government [1]. Core to this development is the increased utilization of digital solutions to achieve organizational change, i.e., digital transformation [2]. There have been numerous accounts of digital transformation in the public sector [3–7]. As highlighted by Mergel et al. [8] this also involves new cooperation modes and an increased emphasis on external utilizations of digital solutions. In other words, digital transformation is changing both the inner workings and the external enactment of government itself.

Digital transformation is executed through a series of digital initiatives in individual organizations. Said initiatives are managed through a traditional process of prioritization, where (often times) projects are proposed, and then compared against other in order

Published by Springer Nature Switzerland AG 2022
R. Krimmer et al. (Eds.): ePart 2022, LNCS 13392, pp. 73–88, 2022.
https://doi.org/10.1007/978-3-031-23213-8_5

to identify which initiatives offer the most value to the organization. In some organizations, this process is managed through a portfolio approach [9], whereas in most public organizations there is no direct aggregate alignment between individual projects and overarching strategic objectives [10].

The initiatives are assessed through often standardized "business cases" [11]. As templates, these documents include assessments of the cost, benefit and risk of each initiative. On the basis of comparing these documents, prioritization groups select the most relevant/critical initiatives that will get a go-ahead and be executed as (often) projects through project charters. Previous studies of the prioritization process have identified biases towards exploitation rather than exploration [12], signaling a risk of the practice being primarily focusing on the short-term demands at the expense of the long-term.

With digital transformation being a strategic priority for an increasing number of public sector organizations [13], digital transformation strategies become increasingly important. As noted by Chanias et al. [14], this particular type of strategy has a tendency to emerge and evolve over time, more resembling a process of strategizing than that of a traditional functional strategy. Perceiving the strategy along those lines, the strategy can be equated with the actual activities conducted in relation to digital, i.e., the total sum of digital initiatives. From this perspective, the business cases and consequent project charters may hold the key to understanding the strategy in use, rather than the strategy in formulation [15].

With the increasing necessity to focus on issues related to sustainability, public sector organizations are more and more adjusting their strategic intent to include the UN based sustainable development goals (SDGs) as concrete objectives in their operations [16]. In other words, digital transformation necessarily needs to converge with the aspiration for the SDGs [17]. Several studies also show that digital transformation is a prerequisite for effective attainment of the SDGs [18–21] and according to a study by Janowski [16], 87% of the SDG targets require the highest level of digital government maturity. Having the highest level of digital government maturity means being capable to transform not only internal organization and external relationships but also to being able to handle complexities and to contextualize and specialize to transform cultures and societies [1].

At the same time, there has been clear concern that the integration of SDGs into public sector operations. In a study of a state-owned enterprise, Montecalvo et al. [22] identify that integrated reporting decreases the risk that sustainability disclosures continue to be merely ceremonial and devoid of actual impact on the organization. Key here is the shift from sustainability being something that the organization reports (output metrics) to it being an integrated element of organizational practice (process metrics). By integrating the SDGs into the very objectives of the organization, i.e., into the very initiatives and investments themselves [23–25], we will be able to assure that each organization actually contributes to the attainment of SDGs and not merely comply with specific accounting standards for external reporting [26]. On the basis of this short rationale, this study answers the following research question: *How are Sustainable Development Goals integrated in digital transformation initiatives?*

The study is operationalized through collecting and analyzing secondary data in the form of a purposive sampling of digital transformation project charters and budgets in a large, Swedish municipality. By extracting each project's objectives and then coding these for SDG linkages, we analyze how the SDGs are integrated into digital transformation initiatives.

The remainder of the paper is organized accordingly: After this brief introduction, we present the previous research on public sector digital transformation, sustainability and the integration between digital transformation and sustainability. This is followed by the method, where we expand on the methodological choices made in the study. After this, we present the results through an overview of the level of integration and a more nuanced analysis of patterns. The paper ends with a discussion where we contrast the results of our study with previous findings, with the intent of presenting clear contributions to research, practice and policy as well as future directions for research.

2 Previous Research

In its broadest sense, digital transformation refers to the utilization of digital technologies in order to drive organizational change. Digital technology challenges existing practices of computing, communication, and connectivity and contributes to a landscape of distributed networks and diversified relationships that extend beyond traditional organizational boundaries [27]. This new reality implies huge challenges for organizations if they are to leverage the benefits of novel, unpredictable, ever-changing technology and related emerging innovations [9, 28].

Digital transformation in the public sector has been broadly conceptualized as various socio-technical arrangements that capture these challenges and contribute to a significant change in service, organization, culture, and relationships to citizens and businesses that substantially increase the scope and scale of public val, ue to citizens [1, 8, 29–31]. The concept of digital transformation implies "holistic efforts to revise core processes and services of organizations beyond traditional digitization efforts [8, p. 12] and thus requires significant citizen centric organizational change, backed up by leadership and cultural change that can leverage digital technology to increase value all actors (e.g. organization, citizens, business, society) [32].

Public sector organizations need to be organized to meet the challenges that new technology and services can provide [33]. The challenges involve understanding, being able to address and to enacting "cultural, organizational and, relational changes" [8] deep iterative transformation over time [34] and multiple socio-technical changes occurring in parallel. Digital transformation processes should not be understood as homogenous entities operating in isolation, but rather as multifaceted change processes that occur through parallel and intertwined activities interplaying with different organizational elements [35]. These processes can therefore never be understood in isolation from changed organizational structure and culture [36].

Traditionally, within the Information systems (IS) field, there has been a unifying vision that the outcome of IS implementation would be to helping organizations to use information systems more effectively [37]. With the fast-growing globalization and following grand challenges characterized by complexities and radical forms of uncertainty beyond local and internal control [38], the agenda has started to being rephrased to include questions of what IS can do to make a better world [37, 39, 40]. In that way, rather than only studying internal effectiveness, there is a growing need in the IS field to focus on "responsible IS" efforts that on the one hand take critical perspectives on what negative effects IS may imply, and on the other hand what good and bad aspects can come out of IS implementation from a broader stakeholder perspective [38].

This development of the IS field is also pushed by global and local calls for common action to achieve different sustainability goals. Agenda 2030 and UN's sustainable development goals (SDGs) are playing an important role in unifying actors at all levels to apply a sustainability "lens" to their activities [41]. Public sector organizations are significant actors in the work of "making the world better" [42] and there is therefore a great need for more research and practice on how SDGs are and can be integrated into public sector organizations [16]. With this background and in combination with the argument that digitalization is a prerequisite for any radical change in organizations [27], we argue that it is important to investigate how SDGs are integrated into digital transformation initiatives.

3 Method and Analysis

This study takes inspiration from phenomenon driven research [43] in that it approaches the areas of digital transformation and sustainability from the phenomenon rather than a specific theory. Instead of deriving the foci of our research directly from previous calls for research, i.e., gap-spotting to build new theory, we utilize a problematization starting from the empirical realm aimed at mapping new constructs to the phenomenon. Through our clinical [44] programmatic [45, 46] research conducted with public sector organizations since 2017 we (the authors of this paper and their research group) have noticed hesitance in the organizations regarding what is being done in terms of sustainability. In parallel with this, we have noticed a similar uncertainty about how digital transformation is executed. It is this phenomenon that is the focus of our study, i.e., the integration of sustainability and digital transformation.

For this study, we worked with a large, Swedish municipality (55,000 employees, 580,000 citizens) with the idea of studying how digital initiatives were directed in terms of the explicit goals of each initiative. With the municipality following a standardized procedure for prioritization and execution of projects, there was a substantial amount of information available for each project in the form of templated project charters and budgets.

We made a purposive sampling [47] of digital initiative portfolios (i.e., collections of projects per organizational entity, see [9] to identify a representative selection of the whole of the municipalities initiatives that would be feasible to approach with our research objective.

Coding was conducted through exerting the effect goals of each project. An effect goal is the measurable objective of the project, i.e., what tangible results the overarching aim of the project is supposed to achieve. In other words, the sum of all effect goals in the city would constitute what the organization is aiming to achieve with their digital initiatives at present, i.e., the digital transformation strategy in [14]. The total empirical material contained 30 projects with 74 effect goals, distributed across ten organizational entities and fourteen portfolios.

Each effect goal was categorized in relation to the UN Sustainable Development Goals (SDGs) with the nuancing of "directly" or "indirectly" related. In terms of the "directly" category, this was when the effect goal was explicitly formulated as a goal that targets one or several SDGs. In terms of the "indirectly" category, this was when SDGs were not explicitly targeted in the formulation of the effect goals. The researchers interpreted that when the goal is achieved, effects of the application of the goal may be related to one or several of the SDGs. In this paper we only analyze the directly related SDGs as they are already integrated into the effect goals and thus directly operative. We see the indirectly related goals as important potentials for the organization to work with, and we will report on these in a future paper. However, as for the purpose of this paper they are too speculative (Table 1).

Table 1. Example of coding from the analysis showing organizational effect goals; directly (DR) and indirectly related (IDR) SDGs and; researchers' motivating comments on how effect goals and SDGs are related.

Effect goal	SDG coding	Comment
"A joint assessment instrument for financial assistance that ensures correct and equal assessment in the social services."	DR SDGs: 16.3, 16.5, 1.4, 10.2, 10.3, 1.2	DR SDGs: The aim of this effect goal is to create an instrument that will contribute to increased legal certainty (16.3) in assessment processes in cases involving financial assistance in the social services. Automation should ensure equal assessment, which can be linked to goals of reduced corruption (16.5), improved economic and social justice (1.4, 10.2, 10.3) and reduced poverty (1.2) (4% in Sweden according to Statistics Sweden, 2020), so that the most vulnerable, less empowered people get the same opportunities for assistance as everyone else

(*continued*)

Table 1. (*continued*)

Effect goal	SDG coding	Comment
	IDR SDGs: 9.1, 16.6, 16.7	IDR SDGs: Indirectly, the instrument could contribute to the creation of sustainable, resilient and inclusive infrastructures, perhaps also cross-border (9.1) and transparent institutions (16.6), that can be critically examined and improved. If this is done in a good way, it could further ensure responsive and inclusive decision-making (16.7)
"Contribute to implementation, and implement, welfare technological solutions within the business areas of ÄO and IFO/FH, which in turn - contributes to the goals of being an attractive employer."	DR SDGs: 9.1	DR SDGs: This effect goal focuses on implementation of welfare technologies which is related to improvement of the digital infrastructure in the city (9.1)
	IDR SDGs: 1.4, 10.2, 10.3, 3.8, 8.5	IDR SDGs: Well-built welfare infrastructure can indirectly increase equal rights to services, technology and financial resources (1.4), accessibility and independence of the user, that in turn contributes to better social inclusion (10.2), reduced discrimination (10.3), and better health care for all (3.8). High technological environments is in this effect goal said to contribute to being an attractive employer, which may mean securing full employment and decent working conditions with equal pay for all (8.5)

(*continued*)

Table 1. *(continued)*

Effect goal	SDG coding	Comment
"To improve the control of lighting in the city and to follow-up of lighting fixtures."	Directly related SDG: 9.1	DR SDGs: With this effect goal the organization will expand the digital and physical infrastructure (9.1) by improving lightening in the city
	IDR SDGs: 11.3, 11.b, 11.4, 11.6, 11.7, 11.1, 11.2, 1.4, 10.2, 10.3	IDR SDGs: It is not clear what the purpose of improving lighting means, but indirectly good lighting in the city can be linked to several SDGs such as including sustainable urbanization (11.3), strategies for sustainable inclusion in the city (11.b), ecologically sustainable city as for example in relation to light pollution (11.4, 11.6), safe green areas (11.7), equal access to safe housing/residential areas (11.1), accessible sustainable transport (11.2) and equal access to the city and its resources, security, crime prevention work, freedom of movement for women and children, etc., i.e. goals linked to socio-economic justice (1.4, 10.2, 10.3)

Once finished with the coding of effect goals and their relation to SDGs, we proceeded to calculate the monetary value of each integration. If a project with a budget of €5M had five effect goals, where only one displayed SDG integration, then the monetary value of said integration was calculated as €1M. With the integration in some cases being distributed over several SDGs, then the monetary value of the individual integration of one SDG was calculated by dividing the value of the integration with the number of integrated SDGs. We calculated the value of integration over both SDGs and portfolios in the sample to be able to analyze how the integration was distributed in the organization (Table 2).

Table 2. Example of calculation of monetary value of integrated SDGs in one project.

Project	Effect goal	Monetary value	SDG	Integration value
1	1	€2M	1, 3, 4	€2M
	2	€2M		0
	3	€2M	1, 2	€2M
	TOT	*€6M*	*1, 2, 3, 4*	*€4M*

The integration value per SDG was calculated (following the above example) according to Table 3.

Table 3. Example of calculation of integration value.

SDG	Integration value
1	(€2M/3) + (€2M/2)
2	(€2M/2)
3	(€2M/3)
4	(€2M/3)

4 Results

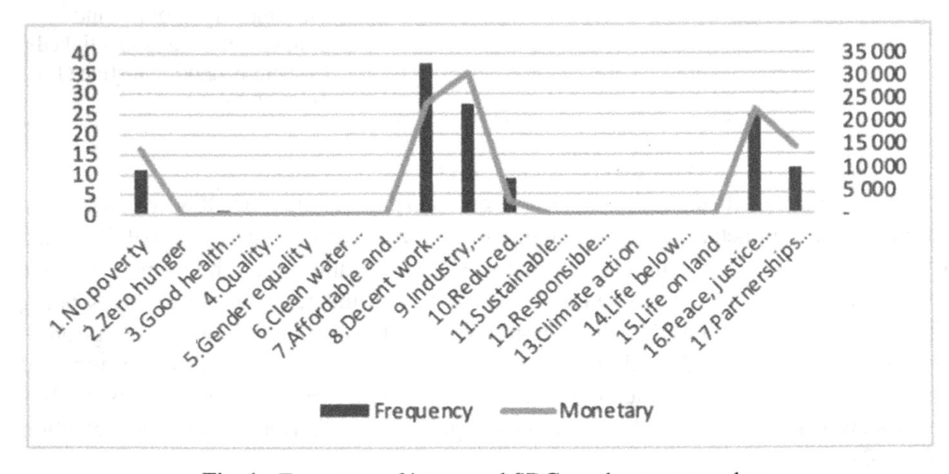

Fig. 1. Frequency of integrated SDGs and monetary value

As seen in Fig. 1, the majority of SDGs are not integrated in the digital transformation initiatives as per the objectives they are to achieve. In other words, we see a clear decoupling of the digital initiatives and the regulated aspirations for increased sustainability. The highest integration is displayed with SDG8, i.e., decent work and economic growth, reverberating the previously acknowledged tendency for digital transformation to be focused on internal efficiency rather than externally toward the citizens in public sector organizations [6]. 62% of the integrated SDGs are directly linked to economic sustainability, with the remaining 38% are linked to social sustainability. None of the integrations are directly linked with environmental sustainability. At the same time, only two out of 102 audited projects did not display at least one integration to SDGs.

Figure 1 also shows the monetary distribution per SDG. In comparison with the frequency, accounting for the financial strength of the integrations leads to a slightly different prioritization of SDGs. Instead of SDG8 being the most prioritized (as per frequency), we see that SDG9 is the SDG with most funds working towards attaining the goal.

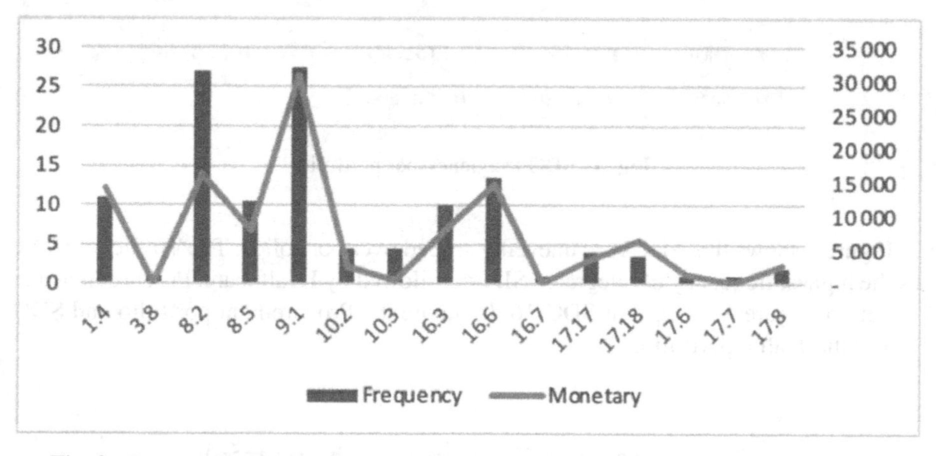

Fig. 2. Frequency of SDG targets (sub-goals) and monetary allocation per SDG target.

When analyzing the SDG targets (sub-goals) (Fig. 2) the dominating targets in terms of the frequency of integration are 8.2 (Achieve higher levels of productivity of economies through diversification, technological upgrading and innovation), 9.1 (Build resilient infrastructure, promote inclusive and sustainable industrialization and foster innovation) and 16.6 (develop effective, accountable and transparent institutions at all levels).

Taking into account the differences in the audited projects in terms of budgets, Fig. 2 also shows the monetary allocation per SDG. The SDG that has the highest priority in terms of monetary allocation is 9.1, followed by 8.2, 16.6 and 1.4. This highlights two things: certain SDGs are not relevant, and, there are prioritizations between the different SDGs.

Comparing the frequency vs monetary, we can see that the relative strength of the levels of funding integrated per SDG decreases the relative value of 8.2. In other words, the money allocated to attaining SDG 8.2 is on par with attaining SDGs 1.4 (To ensure that all men and women, particularly the poor and the vulnerable, have equal rights to economic resources) and 16.6 and only half of 9.1.

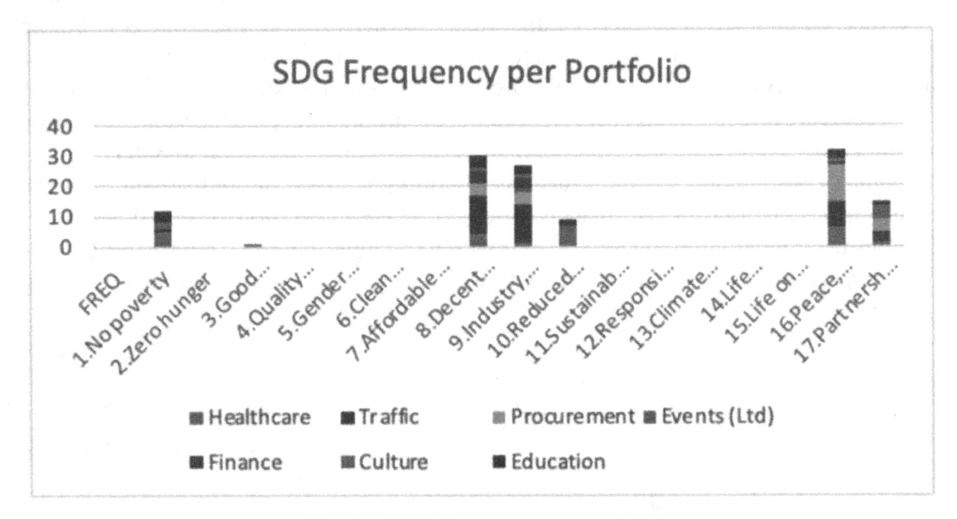

Fig. 3. SDG frequency per portfolio

Figure 3 shows that SDGs are unevenly spread over portfolios. The Traffic portfolio has the highest frequency of integrated SDGs, followed by Healthcare, Procurement and Education. Figure 4 shows that SDG 16 dominate the Procurement portfolio and SDG 8 and 9 the Traffic portfolio.

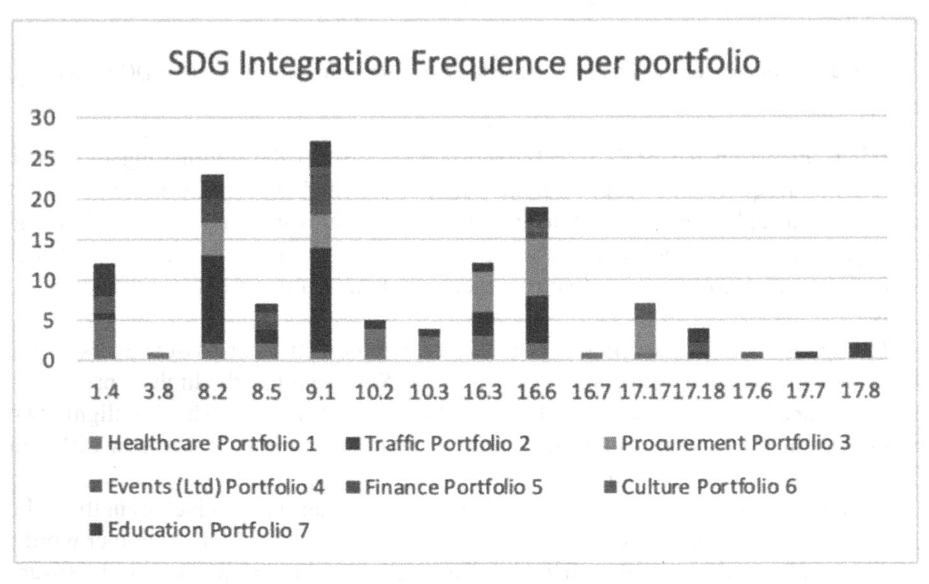

Fig. 4. SDG target integration frequency per portfolio.

Figure 4 shows that SDG targets related to economic productivity (8.2), infrastructure and innovation (9.1) and effective, accountable and transparent institutions (16.6) are most common in the Traffic portfolio, while goals related to equal access to justice for all (16.3) and effective and transparent institutions (16.6) are the most common in the Procurement portfolio. The healthcare portfolio is dominated by targets related to equal rights to economic resources and basic services (1.4), and social, economic and political inclusion of all (10.2, 10.3).

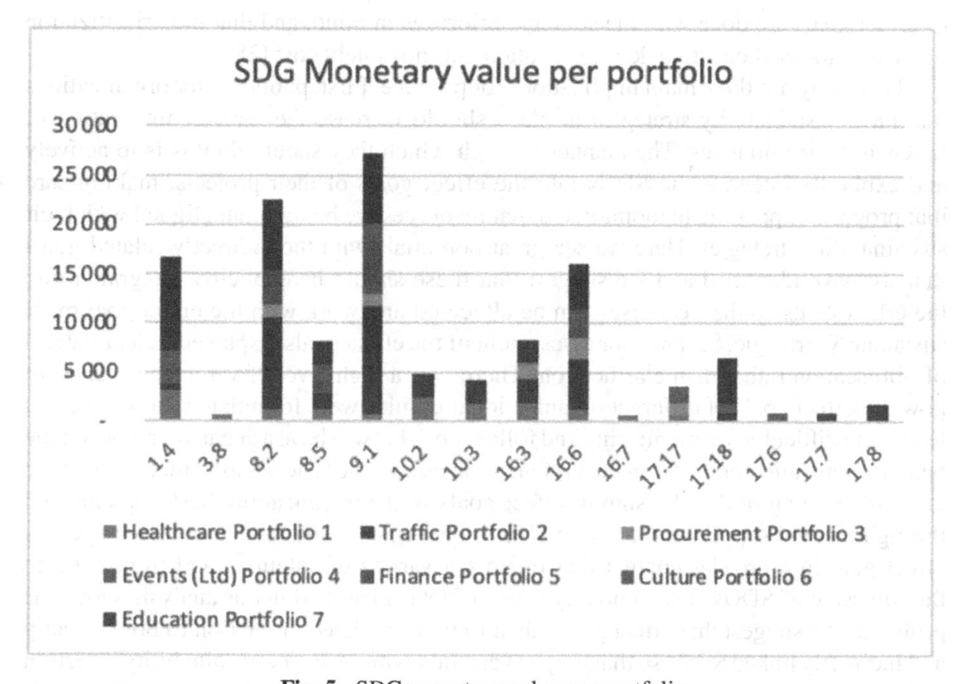

Fig. 5. SDG monetary value per portfolio

Figure 5 shows that the SDG target that gets the most funding is 9.1 and the money is mainly divided between the Traffic, Education and Finance portfolios. The second most funded target is 8.2 and here we see a similar divide of monetary allocation between portfolios.

5 Discussion

This study sets out to answer the research question of how SDGs are integrated into digital transformation initiatives in the public sector. As we found, the overarching level of integration was 98%, i.e., SDGs were integrated in almost all objectives for digital transformation initiatives. At the same time, the main emphasis of integrated SDGs was economic (62%) with zero integration of environmental SDGs. This supports previous findings from studies such as [6, 12] where digital transformation in the public sector is equated with increased internal efficiency, i.e., tightly linked with the SDGs of economic sustainability. The digital transformation initiatives have a tendency to be introvert, not focusing directly on external value such as citizen - or societal - or environmental value. While increased resource utilization efficiency is important for the public sector [48], the dearth of environmental SDGs into digital transformation initiatives is perceived as problematic since strong allocation to digital transformation hence means less allocation to environmental SDGs. The study shows that there are plenty of indirectly linked social and environmental SDGs that organizations could integrate to digital transformation. However for that to be done, practices need to be developed so that the actual formulation of effect goals are done with the strategic priorities in mind, and that the prioritization process is focused on strategic prioritization and not solely cost [28].

This study has three main implications for practice. First, public sector organizations that have sustainability strategies in place should increase the level of integration of SDGs in their initiatives. The manner through which they should do this is to actively and explicitly integrate the SDGs into the effect goals of their projects, making sure that progress is possible to monitor and that resources are being spent aligned with their sustainability strategies. Here we see great potentials with the indirectly related goals that we have identified and we suggest that these should be explicitly integrated into the effect goals so that recourses can be allocated and work with the effect goal more sustainably strategic. Second, our assessment of the effect goals displayed a clear pattern of obfuscation rather than clarification. There was a high level of variance in terms of how the effect goals of different organizational entities were formulated, in some cases leading to difficulties in monitoring and follow up. This will have a negative impact on the benefits realization practices in the organization as a whole. The solution that we propose is a streamlining of the phrasing of effect goals in the organization, best implemented through method support and training. Third, the organization fails to see each specific effect goal in a broader context that mirrors a variety of solutions and in relation to the full scale of SDGs. The indirectly related SDGs identified in our analysis show this problem. We suggest that effect goals need to be formulated in relation to both directly and indirectly linked SDGs so that projects can take a more divers sustainability direction and that relevant recourses can be allocated to the projects. Fourth, we would urge the organizations to implement some sort of support for organization-wide prioritization. At present, there is seldom any overarching monitoring in place, and the complete list of initiatives in a municipality may be non-aligned with the overarching strategy as well as biased in terms of which SDGs are targeted at the different organizational entities. Imposing a new layer of central control for SDG execution would be one way to solve this issue.

We also offer one contribution to policy. With nation states typically having national strategies and visions for SDG compliance in place, the need for national oversight of progress and spend is an important source of data that could help drive better policy [49]. We suggest targeting this through regulating that data on SDG integration should be provided as open data from all organizations in the public sector.

There are three overarching limitations of our study. First, the selection of digital transformation initiatives does not capture the entirety of all the activities related to SDGs in the municipality. We assume that there are multiple projects (e.g. GIS-related projects) directly targeted at SDGs in other parts of the municipality (e.g. urban planning) but given our intention on studying the integration of SDGs in digital transformation initiatives we believe this selection to be justified. At the same time, the findings should not be generalized into an overarching tendency of not focusing on SDGs in the public sector. Second, the decision to focus on the effect goals of the projects rather than the overarching aim may be perceived as not amply capturing the nuances of SDGs in the projects as such. Our justification for the focus on effect goals is two-fold. First, these objectives should (according to the project methodology in the municipality) be a true operationalization of the project as such, i.e., the overarching project aim will be fulfilled through achieving the effect objectives. Second, these objectives are both the basis for prioritization and for monitoring/ex-post evaluation of the project. In other words, if something is not expressed explicitly in the effect goal, it should not be prioritized and will not constitute a desirable aim. The third limitation concerns the difficulty with transferring findings from one specific institutional environment such as a Swedish municipality into another institutional environment such as a German agency [50].

We identify two avenues for future research from our study. First, we would suggest research follow up on the limited integration of SDGs in digital transformation initiatives. As noted by Janowski [16], digital government is instrumental for the achievement of a vast majority of the SDG targets. Hence, we believe that future research should extend our limited study to conduct a comparative, larger study of both central and local government organizations of various sizes and functions. The findings of this will both be able to test the generalizability of our findings, as well as identify contingencies for which factors impact SDG integration in digital transformation initiatives. Second, we see the need for a qualitative, follow-up study where the practices of SDG integration is in focus. This study should select respondents that are directly involved in not only the creation and prioritization of initiatives, but also in the factual management of the projects and subsequent portfolios. This study is important to understand how effect goals, including integrated SDGs, are applied withing projects and how fulfilment of goals are assessed.

6 Conclusion

This study aimed to answer the research question of how SDGs are integrated into digital transformation initiatives in the public sector. Through analysis of secondary data in the form of project charters and budgets, we assessed the level of integration in multiple portfolios in a large, Swedish municipality. This study finds that there is a generally low level of SDG integration in digital transformation initiatives, and present patterns of which SDGs are integrated in which entity in the organization. The implications of our findings point to the need for further studies and changes in practice, where public sector organizations should work more structured and purposive with aligning current digital transformation initiatives with SDGs.

References

1. Janowski, T.: Digital government evolution: from transformation to contextualization. Gov. Inf. Q. **32**, 221–236 (2015). https://doi.org/10.1016/j.giq.2015.07.001

2. Hanelt, A., Bohnsack, R., Marz, D., Marante, C.A.: A systematic review of the literature on digital transformation: insights and implications for strategy and organizational change. J. Manag. Stud. **58**, 1159–1197 (2021). https://doi.org/10.1111/joms.12639
3. Andersen, K.N., Lee, J., Henriksen, H.Z.: Digital sclerosis? Wind of change for government and the employees. Digit. Gov. Res. Pract. **1**, 1–14 (2020). https://doi.org/10.1145/3360000
4. Janssen, M., van der Voort, H.: Adaptive governance: towards a stable, accountable and responsive government. Gov. Inform. Q. **33**, 1–5 (2016). https://doi.org/10.1016/j.giq.2016.02.003
5. Lindgren, I., Madsen, C.Ø., Hofmann, S., Melin, U.: Close encounters of the digital kind: a research agenda for the digitalization of public services. Gov. Inf. Q. **36**, 427–436 (2019). https://doi.org/10.1016/j.giq.2019.03.002
6. Magnusson, J., Päivärinta, T., Koutsikouri, D.: Digital ambidexterity in the public sector: empirical evidence of a bias in balancing practices. Transforming Gov. People Process Policy **15**, 59–79 (2020). https://doi.org/10.1108/tg-02-2020-0028
7. Mergel, I.: Agile innovation management in government: a research agenda. Gov. Inf. Q. **33**, 516–523 (2016). https://doi.org/10.1016/j.giq.2016.07.004
8. Mergel, I., Edelmann, N., Haug, N.: Defining digital transformation: results from expert interviews. Gov. Inf. Q. **36**, 101385 (2019). https://doi.org/10.1016/j.giq.2019.06.002
9. Lappi, T.M., Aaltonen, K., Kujala, J.: Project governance and portfolio management in government digitalization. Transforming Gov. People Process Policy **13**, 159–196 (2019). https://doi.org/10.1108/tg-11-2018-0068
10. Bryson, et al.: pdf, n.d. (2020)
11. Nielsen, P.A., Persson, J.S.: Useful business cases: value creation in IS projects. Eur. J. Inf. Syst. **26**, 66–83 (2017). https://doi.org/10.1057/s41303-016-0026-x
12. Magnusson, J., Koutsikouri, D., Päivärinta, T.: Efficiency creep and shadow innovation: enacting ambidextrous IT Governance in the public sector. Eur. J. Inf. Syst. **29**, 1–21 (2020). https://doi.org/10.1080/0960085x.2020.1740617
13. Wilson, C., Mergel, I.: Overcoming barriers to digital government mapping the strategies of digital champions. Gov. Inf. Q. 101681 (2022). https://doi.org/10.1016/j.giq.2022.101681
14. Chanias, S., Myers, M.D., Hess, T.: Digital transformation strategy making in pre-digital organizations: the case of a financial services provider. J. Strateg. Inf. Syst. **28**, 17–33 (2019). https://doi.org/10.1016/j.jsis.2018.11.003
15. Menz, M., et al.: Corporate strategy and the theory of the firm in the digital age. J. Manag. Stud. **58**, 1695–1720 (2021). https://doi.org/10.1111/joms.12760
16. Janowski, T.: Implementing sustainable development goals with digital government – aspiration-capacity gap. Gov. Inf. Q. **33**, 603–613 (2016). https://doi.org/10.1016/j.giq.2016.12.001
17. Del Río Castro, G., Fernández, M.C.G., Colsa, Á.U.: Unleashing the convergence amid digitalization and sustainability towards pursuing the Sustainable Development Goals (SDGs): a holistic review. J. Clean. Prod. **280**, 122204 (2020). https://doi.org/10.1016/j.jclepro.2020.122204
18. Adams, R., Kewell, B., Parry, G.: Blockchain for good? Digital ledger technology and sustainable development goals. In: Leal Filho, W., Marans, R.W., Callewaert, J. (eds.) Handbook of Sustainability and Social Science Research. WSS, pp. 127–140. Springer, Cham (2018). https://doi.org/10.1007/978-3-319-67122-2_7
19. ElMassah, S., Mohieldin, M.: Digital transformation and localizing the Sustainable Development Goals (SDGs). Ecol. Econ. **169**, 106490 (2020). https://doi.org/10.1016/j.ecolecon.2019.106490
20. Hoosain, M.S., Paul, B.S., Ramakrishna, S.: The impact of 4IR digital technologies and circular thinking on the United Nations sustainable development goals. Sustainability **12**, 10143 (2020). https://doi.org/10.3390/su122310143

21. Marcovecchio, I., Thinyane, M., Estevez, E., Janowski, T.: Digital government as implementation means for sustainable development goals. Int. J. Public Adm. Digit. Age (IJPADA) **6**, 1–22 (2019). https://doi.org/10.4018/ijpada.2019070101
22. Montecalvo, M., Farneti, F., de Villiers, C.: The potential of integrated reporting to enhance sustainability reporting in the public sector. Public Money Manag. **38**, 365–374 (2018). https://doi.org/10.1080/09540962.2018.1477675
23. Forestier, O., Kim, R.E.: Cherry-picking the Sustainable Development Goals: goal prioritization by national governments and implications for global governance. Sustain. Dev. **28**, 1269–1278 (2020). https://doi.org/10.1002/sd.2082
24. Gustafsson, S., Ivner, J.: Implementing the global sustainable goals (SDGs) into municipal strategies applying an integrated approach. In: Leal Filho, W. (ed.) Handbook of Sustainability Science and Research. WSS, pp. 301–316. Springer, Cham (2018). https://doi.org/10.1007/978-3-319-63007-6_18
25. Raub, S.P., Martin-Rios, C.: "Think sustainable, act local" – a stakeholder-filter-model for translating SDGs into sustainability initiatives with local impact. Int. J. Contemp. Hosp. Manag. **31**, 2428–2447 (2019). https://doi.org/10.1108/ijchm-06-2018-0453
26. Flower, J.: The international integrated reporting council: a story of failure. Crit. Perspect. Accoun. **27**, 1–17 (2015). https://doi.org/10.1016/j.cpa.2014.07.002
27. Vial, G.: Understanding digital transformation: a review and a research agenda. J. Strateg. Inf. Syst. **28**, 118–144 (2019). https://doi.org/10.1016/j.jsis.2019.01.003
28. Lappi, T., Aaltonen, K.: Project governance in public sector agile software projects. Int. J. Manag. Proj. Bus. **10**(2), 263 (2017)
29. Tangi, L., Janssen, M., Benedetti, M., Noci, G.: Digital government transformation: a structural equation modelling analysis of driving and impeding factors. Int. J. Inf. Manag. **60**, 102356 (2021)
30. Lindgren, I., van Veenstra, A.F.: Digital government transformation: a case illustrating public e-service development as part of public sector transformation. In: Proceedings of the 19th Annual International Conference on Digital Government Research: Governance in the Data Age, pp. 1–6 (2018)
31. Stoker, G.: Public value management: a new narrative for networked governance? Am. Rev. Public Adm. **36**(1), 41–57 (2006)
32. Vallo Hult, H., Byström, K.: Challenges to learning and leading the digital workplace. Stud. Continuing Educ. **44**(3), 460–474 (2022). https://doi.org/10.1080/0158037X.2021.1879038
33. Ashaye, O.R., Irani, Z.: The role of stakeholders in the effective use of e-government resources in public services. Int. J. Inf. Manag. **49**, 253–270 (2019)
34. Pittaway, J.J., Montazemi, A.R.: Know-how to lead digital transformation: the case of local governments. Gov. Inf. Q. **37**, 101474 (2020). https://doi.org/10.1016/j.giq.2020.101474
35. Gong, Y., Yang, J., Shi, X.: Towards a comprehensive understanding of digital transformation in government: analysis of flexibility and enterprise architecture. Gov. Inf. Q. **37**(3), 101487 (2020)
36. Pedersen, K.: E-government transformations: challenges and strategies. Transforming Gov. People Process Policy **12**(1), 84–109 (2018)
37. Walsham, G.: Are we making a better world with ICTs? Reflections on a future agenda for the IS field. J. Inf. Technol. **27**(2), 87–93 (2012)
38. Pan, S.L., Zhang, S.: From fighting COVID-19 pandemic to tackling sustainable development goals: an opportunity for responsible information systems research. Int. J. Inf. Manag. **55**, 102196 (2020). https://doi.org/10.1016/j.ijinfomgt.2020.102196
39. Gil-Garcia, J.R., Zhang, J., Puron-Cid, G.: Conceptualizing smartness in government: an integrative and multi-dimensional view. Gov. Inf. Q. **33**, 524–534 (2016). https://doi.org/10.1016/j.giq.2016.03.002

40. Wynn, E., Vallo Hult, H.: Qualitative and Critical Research in Information Systems and Human-Computer Interaction: Divergent and Convergent Paths. Found. Trends Inf. Syst. **3**(1–2), 1–233 (2019)
41. United Nations: The 17 goals (2022). https://sdgs.un.org/goals. Accessed 18 Mar 2022
42. Regeringskansliet: Agenda 2030 och de globala målen för hållbar utveckling (2022). https://www.regeringen.se/regeringens-politik/globala-malen-och-agenda-2030/17-globala-mal-for-hallbar-utveckling/. Accessed 08 Mar 2022
43. Schwarz, G., Stensaker, I.: Time to take off the theoretical straightjacket and (re-)introduce phenomenon-driven research. J. Appl. Behav. Sci. **50**, 478–501 (2014). https://doi.org/10.1177/0021886314549919
44. Schein, E.H.: Clinical inquiry/research. In: The SAGE Handbook of Action Research: Participative Inquiry and Practice, pp. 266–279. Sage, New York (2008)
45. Burton-Jones, A.: Minimizing method bias through programmatic research. MIS Q. **33**, 445 (2009). https://doi.org/10.2307/20650304
46. Covin, J.G., McMullen, J.S.: Programmatic research and the case for designing and publishing from rich, multifaceted datasets: issues and recommendations. J. Bus. Res. **101**, 40–46 (2019). https://doi.org/10.1016/j.jbusres.2019.04.012
47. Kopmann, J., Kock, A., Killen, C.P., Gemünden, H.G.: The role of project portfolio management in fostering both deliberate and emergent strategy. Int. J. Proj. Manag. **35**, 557–570 (2017). https://doi.org/10.1016/j.ijproman.2017.02.011
48. Wassmer, U., Li, S., Madhok, A.: Resource ambidexterity through alliance portfolios and firm performance. Strateg. Manag. J. **38**, 384–394 (2017). https://doi.org/10.1002/smj.2488
49. van Veenstra, A.F., Kotterink, B.: Data-driven policy making: the policy lab approach. In: Parycek, P., et al. (eds.) ePart 2017. LNCS, vol. 10429, pp. 100–111. Springer, Cham (2017). https://doi.org/10.1007/978-3-319-64322-9_9
50. Bannister, F.: The curse of the benchmark: an assessment of the validity and value of e-government comparisons. Int. Rev. Adm. Sci. **73**, 171–188 (2007). https://doi.org/10.1177/0020852307077959

Digital and Social Media

Sharing, Cooperation or Collective Action? A Research Agenda for Online Interaction in Digital Global Governance

Tove Engvall(✉), Leif Skiftenes Flak, and Øystein Sæbø

University of Agder, Postboks 422, 4604 Kristiansand, Norway
{tove.engvall,leif.flak,oystein.sabo}@uia.no

Abstract. Digital technologies are increasingly used to support governance at the global level. However, the global level has received very little attention in digital governance research. Global governance differs from national governance contexts in that it does not have a central government with authority of enforcing decisions. Consequently, as engagement of stakeholders is vital for taking appropriate action, possibilities and challenges in using digital technologies to facilitate response to common challenges should be further investigated. To address this shortcoming, we explore how digital technologies and online communities can leverage participation and co-production in the context of global governance. Based on an existing classification of online interaction (sharing, cooperation, collective action) we suggest a research agenda that can move the knowledge front related to online interactions in global governance contexts.

Keywords: Digital governance · Digital global governance · Online communities · Co-creation · Citizen engagement · Sustainability · UN global stocktake

1 Introduction

Digitalization transforms the way public sector organizations work and interact, both within and between organizations, as well as with external stakeholders. Digital technologies offer new venues for political discussions [1] and for organizations to interact with stakeholders [2]. With societal challenges being increasingly global in character, the need for global coordination and response increases. A pertinent question is then how digital technologies could contribute to enhance global governance as a response to global challenges. 'Digital governance' is the research field that investigates the use of digital technologies in governance structures and processes, and it has evolved through the concepts of eGovernment, eGovernance and digital governance. The term 'Digital global governance' refers to the use of digital technologies in global governance structures and processes. Unfortunately, little research in the digital governance field addresses the global governance level but tends to focus on either the national or municipal level. International studies are primarily national comparisons of eGovernment development

Published by Springer Nature Switzerland AG 2022
R. Krimmer et al. (Eds.): ePart 2022, LNCS 13392, pp. 91–106, 2022.
https://doi.org/10.1007/978-3-031-23213-8_6

in different countries [3–6], where common themes are digital divide [7] and diffusion of digital governance [8]. A few notable exceptions exist on initiatives of global character focusing on global digital citizenship [9], global ICT programs [10] and global civil society networks [11]. However, studies on digitalization of global governance structures and processes seem to be largely missing.

Triggered by the Covid pandemic, online tools have increasingly been used in global governance processes. For instance, the annual meeting in Glasgow 2021 of the supreme decision-making body of the Climate Convention, COP (Conference of the Parties), used an online platform to increase possibilities for participation. Some of the sessions were streamed to the public, and multiple social media channels used [12]. In May–June 2021, the UN Climate Change subsidiary bodies sessions were carried out fully online, including dialogues and discussions to prepare for negotiations [13]. It has further been argued that digitalization may enable a larger change of the climate governance process, which has been requested by various stakeholders. The critique of the current process includes mistrust, power imbalances and polarization, as well as insufficient outcomes that fail to adequately address the climate change challenge [14].

Governance can be understood as the steering of society according to common goals, through collective action [15]. Governance in a global context differs from governance at national level. An important difference is that governance beyond nation states lacks a central authority of a government, which has a legitimate use of force [16]. Instead, common agreements, consensus, and trust are significant. Global governance engages multiple actors with different roles. Stakeholder interaction, both within an organization and with external stakeholders, is important from a democratic perspective, regarding both a capability to make agreements, consider various perspectives, and collaborate. Stakeholder engagement also has an important role in strengthening implementation capability of international agreements.

Research on Online Communities has shown how people use digital technologies to organize collective action in the online environment, characterized by not having a traditional organization with a central authority [17]. In that sense, it has commonalities with consensus-based global governance. We argue that facilitation of collective action strengthens the global community's capability to respond to common societal challenges, and experiences from research in online communities on collective action can inform how digital technologies can be used to enhance responsiveness to global challenges. Based on a classification of different degrees of involvement of stakeholders; information sharing, cooperation, and collective action [18], this research note develops a research agenda for online interactions in global governance settings. Research notes often follows a less strict paper outline than research papers and are typically used to advance new ideas or, as in our case, research agendas. Thus, research notes are often less reliant on formal research methods but equally reliant on quality through polemic clarity and rhetoric rigor [19].

We use climate governance as an illustrative example to demonstrate the relevance of the research agenda. The guiding research question for this research agenda is: how could online interaction be developed in global governance and what research questions ought to be considered?

2 Theoretical Foundation

Our conceptual framework draws on research on online communities and situates it in the field of digital governance research. The digital governance domain addresses digitalization of governance structures and processes. Research on online communities provides understanding of socio-political engagement and interaction in online contexts. We suggest that bridging these strands of research offers novel ways to generate knowledge to better understand what happens when governance processes shift from physical to virtual arenas.

2.1 Digital Governance

Governance can be defined as "*The process of steering society and the economy through collective action and in accordance with common goals*" [15]. Global governance means that authority is exercised across national borders, and justified by transnational problems or global common goods [20]. A crucial way in which global governance differs from national governance is that it does not have a central government with authority to enforce decisions [16]. Therefore, engagement of stakeholders (both governments and other stakeholders) to take appropriate action for the benefit of the common good is crucial.

Digital governance can be defined as "*digital technology ingrained in structures or processes of governance and their reciprocal relationships with governance objectives and normative values. Digital governance includes the utilization of digital capabilities and involves a transformation of structures, processes or normative values*" [21]. Transformations of governance can be structural and normative, where structural transformations are changes of structures and processes, and normative transformations are related to the qualities of governance, such as transparency, accountability, efficiency and effectiveness [22]. 'Digital global governance' is this understanding of digital governance applied to global governance.

Digital governance is increasingly ingrained in modernization strategies in the public sector, to improve processes and to create public value. EU's agenda towards evidence based and data driven policy making is for instance argued to improve policy processes and decision making, and support collaborative working processes with participation of stakeholders [23, 24].

Digital governance has emerged over time, also conceptually. While eGovernment primarily focused on digitalization of public administration, eGovernance is a broader concept which also transforms various relations (such as Government - Citizen (G2C), Government – Businesses (G2B), Government – Government (G2G)). Digital governance is based on this, with slightly more emphasis on computational capabilities, including data analysis, modeling and visualization [21]. Global governance includes both G2G, G2B and G2C relations. Digitalization increasingly transforms governance in various ways, and it is argued that broader questions of governance in the digital era [25] and the integration of digital technologies in policy processes [26, 27] are needed.

Digital governance is often argued to contribute to increased transparency, good governance and to enable new forms of participation [21]. Public sector organizations that are traditionally recognized as being hierarchical and bureaucratic, are now opening up for various forms of broader participation, within organizations and with external stakeholders, with the aim to improve public value creation. Digital technologies can support co-production, sharing of information and provide tools and methods for citizen-government interaction. However, a move towards public participation and co-production requires both technological, organizational, cultural and competence-related changes [28]. To transform into more participatory models, fostering a participatory culture is key. A participatory culture is characterized by participants experiencing a connection with others and that their contributions matter. The concept of participatory culture elucidates the shared social practice and culture of engaging, participating, and contributing to a community. A participatory culture may encourage empowerment, civic engagement and improve legitimacy of public sector organizations. Digital technologies can enable new ways for stakeholders to engage, participate, contribute, and interact.

An advanced form of participation is co-production. Co-production and collaborative innovation are processes where organizations work with external stakeholders to achieve some outcome together. "Collaborative innovation is a process of creative problem solving through which relevant and affected actors work together across formal institutional boundaries to develop and implement innovative solutions" [28]. The role of the public administration is in this context to facilitate co-production. In order for public sector organizations to develop a participatory culture, they need to "establish a range of processes, infrastructure and policies that ensure that stakeholders can participate" [28], and external stakeholders need to develop skills and capabilities to participate meaningfully.

However, participation may not always lead to desired outcomes, but rather sometimes to destruction of value. It may involve conflicts, marginalization of certain actors and domination of others, power imbalances, misinformation, and misuse of public resources, caused by either internal or external barriers and challenges [28]. Concerns have been raised about the relationship between social media, political polarization, and political disinformation, and its democratic effects. A part of this complexity are automated online propaganda bots [29]. More research is suggested on the role of public organizations and also whether anticipated effects of digitalization are actualized [28].

In general, digitalization and digital governance have associated risks and challenges, such as digital divide, misinformation, challenges of trust, illicit surveillance, cyber security issues and information overload [21]. In a participatory environment, the vulnerability to these risks may increase. A holistic approach that considers both possibilities and risks with digitalization ought to be acquired to deliberately design solutions for appropriate levels of online interaction.

2.2 Online Communities

The concept of community relates to the social relationship of members of a closed area of people, characterized by a defined size, membership and geographical boundaries as well as shared beliefs, values and historical experiences [30]. Weber argues that social action is based on common membership in a community, defined on the orientation of mutual attitudes of individuals' subjective awareness of specific situations [31].

Online communities (OC), is the persistent collections of people with common interests whose primary method of communication is the Internet, typically by the use of social media [32]. OC offers new channels for organizations to connect with stakeholders and provide venues for political and social discussion [1]. OCs are being increasingly explored by organizations for a variety of purposes, including managing relations with customers and partners [33, 34], cooperating on knowledge generation [17, 35] and sharing information of public interest [36]. Unlike traditional communities, pre-existing social ties and material benefits for contributions are weak or non-existent in online groups [37], allowing for broader organization-wide online sharing [35] to become more flexible and fluid than in traditional communities [17].

With the introduction of digital technologies, the transaction costs of communication drops, making it easier for people to get together and organize [18]. IT changes and supplants the role of hierarchy into networks [38], characterized by being organized based on strength and competence, relational communication patterns, conflicts resolved through norms, flexibility, commitment based on mutual benefits and relationship governed by interdependencies [39].

The management of online communities may be influenced by complexity regarding size, diversity and the type of work being created. Work - related activities often foster interpersonal ties, whereas groups focusing on non-work-activities such as political causes [40, 41] often share a common purpose and are likely to behave differently than online groups organized around work - related topics [37]. Ren et al. [2] found that identity-based features needed in online communities sharing common purpose, had stronger effects than bond-based features needed in work - related online communities, arguing that more research is needed to explore these differences.

Shirky [18] provides a simplified, yet illustrative classification of various forms of group undertakings in electronic networks by proposing a three-step ladder of online group interaction.

Sharing represents the easiest group of compilation with fewest demands on the participants. Sharing platforms allow everyone to share and receive in a "take it or leave it fashion" which allows for freedom for individuals and few complications for the group's life, where the group is mainly the aggregate of participants [18]. Digital tools may be used for knowingly sharing for instance pictures, messages, or work files with others.

Cooperation is the next rung on the ladder, representing a more complex situation than simply sharing, since it involves changing behavior to synchronize with others. Cooperation creates group identity since you know who you are cooperating with. Conversation represents a simple form of cooperation, either face to face or by the various use of ICT. While the increased sense of community using online tools should be seen as a positive effect of cooperation, it is also difficult to keep online communication targeted around a specific topic. As a result, some sets of common agreed mechanisms are often needed. Collaborative production/co-production represents a more involved form of cooperation, where no individual can take credit for the results of the process, which could not come into being without the participation of many. Here (unlike sharing) some collective decisions must be made to negotiate about the results, for instance the resulting Wikipedia article.

Collective Action represents the more advanced kind of group efforts. Here, shared responsibility is of critical importance to link individual user identity with the identity of the group, which holds the power in making group decisions which are binding for all individual members. As argued by Shirky [18]: For a group to take collective action, it must have some shared vision strong enough to bind the group together, despite periodic decisions that will inevitably displease at least some members. For this reason, collective action is harder to arrange than information sharing or collaborative creation.

The more common collective action problem is the "tragedy of the commons", wherein individuals have an incentive to damage the collective good. For instance, when all countries agree that CO_2 emissions need to be reduced, but every individual country may benefit from not reducing their own emissions. Therefore, rules are needed, making collective action harder to arrange than sharing or collaborative creation (cooperation). While ubiquitous access to communication tools makes it easy to initiate various forms for group activities, the main challenge is to use tools to promote collaborative collective actions to avoid the adverse outcomes of independent actions [42].

Below is a table that explains the different levels of group interaction based on Shirky [18] (Table 1):

Table 1. Level of online interaction

	Outcome	Level of interactions	Level of coordination/rules
Sharing	Sharing of content among a huge (unrestricted?) number of individuals	Limited need for channels to distribute content	Providing access for everyone to share content
Cooperation	Content produced because of the efforts made by many	Interactions needed to support conversation, negotiations and collective decisions resulting in an agreed outcome	Common agreed rules on how to navigate from individual ideas to a joint result
Collective Action	Collective decisions binding for all individual members	Interactions needed to agree and maintain a shared vision strong enough to bind members being displeased with some decisions	Rules to reduce the problem of the "tragedy of the commons"

2.3 The Example of the Global Stocktake in Global Climate Governance

To illustrate the different levels of interaction, the Global Stocktake of progress towards the goals in the Paris Agreement is selected as an example. It was selected because it has a process that illustrates different levels of interaction among participants.

The Paris Agreement is the most recent international agreement on climate change, adopted within the United Nations Framework Convention on Climate Change (UNFCCC). The Paris Agreement has established common goals on climate governance (on emission reduction, climate adaptation and means of implementation in terms of finance and technology) [43]. Every fifth year (the first time 2021–2023), a Global Stocktake is carried out, where collective progress towards the goal in the Paris Agreement is analyzed and assessed, and further needs for action is identified. The Global Stocktake is carried out in three phases; information collection and synthetization, technical assessment, and negotiation and adoption of a declaration [44]. The UNFCCC secretariat provides technical expertise and organizational support to the process. The secretariat also hosts the registries and systems managing the reports that countries regularly submit due to reporting requirements in the Paris Agreement [45]. In this paper, the Global Stocktake is used to exemplify the usefulness and relevance of the agenda in the context of global climate governance. The research agenda is based on the three levels of interaction in online communities, namely sharing, cooperation and collective action, as outlined by Shirky [18].

3 Research Agenda for Online Interaction in Digital Global Governance

This section outlines a research agenda for online interaction in digital global governance processes. Above we have discussed digital global governance, the increasing importance of digital tools in a governance context, possibilities with online participation and the need to better understand such development by exploring the role of online communities. Here, we isolate recurring themes and develop them into a more general research agenda for online interactions in digital global governance.

3.1 Sharing in Digital Global Governance

Information sharing means that information is shared among an extensive number of individuals [18]. This includes not only dissemination activities but also the collection of information from various stakeholders, to support the data-information workflow.

A key concern within our running example of climate governance is to collect, organize and disseminate information. Based on the global challenge of climate change, information is reported to the UNFCCC by countries. The potential outcome is to provide knowledge on the global status and a common basis for identifying needs for action, decision making and a shared vision. There can also be forums for dialogues with external stakeholders, and means to provide input, for instance related to high level meetings. Information sharing is needed to develop common awareness and understanding of topics. A challenge is that countries have different capacities and conditions for collecting and reporting information according to reporting requirements. Another challenge is to create meaning in the large volumes of information and to make it understandable to various stakeholders. As the information is used to inform governance, it is crucial that it is of high quality and trustworthy.

As discussed above, information sharing represents the more basic level of interaction within online communities. Hence, activities here are assumingly less controversial than activities related to cooperation and collective action. Still more research is needed to better understand both the interaction and the coordination mechanism for successful sharing of information.

Digital tools allow for almost unlimited collection and dissemination of information, from various stakeholders and sources, and a question is who is considered a legitimate provider of information. A research question relating to the level of interaction is ***how technology can facilitate the collection of high-quality information from appropriate stakeholders.***

This further relate to the need for more research at the level of coordination of information management, to better understand ***how technology influences the quality, flow, and presentation of information to various stakeholders*** within the area of digital global governance. To have value, the information must be standardized and comparable to enable synthetization and coordination at global level, also over time; meet certain quality requirements; and be organized and presented in ways that inspire action by various stakeholders.

Referring to the example of the Global Stocktake, countries report regularly national information according to standards and reporting requirements, including greenhouse gas emissions, commitments, and measures on climate action, which is accessible on the UNFCCC website. These reports are the foundation for the synthesis reports that form the input to the technical assessments in the Global Stocktake process. Information for the Global Stocktake is gathered on a special side of the UNFCCC website. A digital submission portal is also set up for external stakeholders to provide input to the Global Stocktake [44, 46]. A great challenge is to organize the massive amounts of information from countries all over the world, and to present the information in ways that are understandable and engaging to stakeholders. Yet another challenge is to provide means to organize external stakeholders' views in the Global Stocktake process.

3.2 Cooperation in Digital Global Governance

Information sharing activities are necessary pre-requisites for the next level of group interactions, the cooperation activities. Cooperation is important in order to have conversations around a problem, current status and needs for action, to identify solutions, and establish a common ground for decision making, and to establish a common identity and a sense of community [18]. The potential outcome of cooperation in a global governance setting is an agreed knowledge status on a topic, or on progress towards an agreed global goal. In the case of the Global Stocktake, this would mean an agreement on progress and needs for action towards the goals in the Paris Agreement.

Cooperation activities may require more profound challenges to organization than information sharing activities since the main goal is to gain agreements. More research is needed to understand how such activities could be organized within digital global governance at both the level of interaction and the level of coordination through information management.

At the interaction level, focusing on how to organize conversations, negotiations and sharing of views to guide collective decisions, more research is needed to understand ***the role of technology in synthesizing and leveraging actionable information.*** A key concern is to organize the online discourse respecting the need for a debate characterized by rationality (logical claims and arguments), relevance (stick to the topic), equality (adequate opportunities to participate), reciprocity (listening to each other's arguments) and politeness (showing respect) [47]. A key consideration is how to use the technology wisely to be able to identify useful content within a (potentially) huge amount of information being produced by various stakeholders.

Procedures and rules are clearly needed guiding the process of reaching agreement. Hence a main research topic relates to ***the relationships between rules and regulations, digital solutions and consensus forming.*** Questions of concern involve issues like whom to include at what level, how to resolve disagreements, who has authority to make decisions, how and when to open and close the processes needed to come to an agreement, and how to manage informal power imbalances. Research should also investigate ***how technological, organizational, cultural and competence-related factors influence cooperation*** and active engagement, participation and contribution, where participants feel that they have a connection with others in the community and that their contribution matters [28].

In the example of the Global Stocktake, there is a procedure for the technical assessment in the Global Stocktake, with decisions on what information that will be considered and how information input can be provided [46]. A challenge is the large volume of information that should be synthesized to a global picture, based on what is reported by countries. This is used in the technical assessments and should be communicated in a way that inspires confidence among participants to take appropriate action. Another challenge is further to facilitate the technical assessment dialogues, and to synthesize the outcome of those dialogues into a synthesis report that participants agree on. This relates to the question of how technology can be used to synthesize and leverage actionable information. There could as well be potentially very large volumes of information submitted by other stakeholders as input to the Global Stocktake. A research question is ***how technology can be used to synthesize information from external stakeholders and include it in a meaningful way in the process.***

3.3 Collective Action in Digital Global Governance

Collective action, where people create something together, share responsibility and make decisions that are binding for all participants [18], represents the most advanced level of group interactions within online communities. The potential outcome in global governance is collective decisions that are binding for all individual members.

Research related to the level of interaction now includes the exploration of ***the roles of technology in decision making processes.***

The previous levels of information sharing and cooperation activities are necessary to succeed with collective action. The levels of interaction can be viewed chronologically.

First, there is a need for shared knowledge on a topic, then dialogues to establish a shared understanding on the needs for action is required, which lays the ground for collaborative decision-making based on a shared vision and goals. Research is needed to better understand how technology could support the voting procedures, and to support accountability and evaluation of implementation efforts. These questions relate directly to the level of coordination, where a key research question is ***what the relationship is between technology and trust in the negotiation process, and how to mitigate decisions resulting in "tragedy of the commons".*** Further on, a relevant question is ***what the role technology could have in processes of accountability and follow up on adopted decisions.***

Research should further investigate ***the relationships between technology and co-production*** that enhance implementation capability. Co-production means that organizations work with external stakeholders to together achieve some outcome [28]. In a global governance setting, this could include both collaboration between governments, but also between governments and other stakeholders. Research is suggested to investigate how co-production and collaborative innovation as a process of creative problem solving through collaboration could be facilitated. In order to do that, processes, policies, technologies and skills and competencies required by involved participants should be developed [28]. One prominent issue is the matter of power balances between stakeholders and the concept of salient stakeholders, i.e. who has influence in the process [48]. Currently, the UNFCCC process has been criticized for power imbalances, also with concerns that technology might serve to consolidate existing power structures rather than challenge these. However, it is also argued that digitalization may be a means to change such power imbalances [14]. It is further important to identify risks for co-destruction, which can be caused by conflicts, marginalization of some actors, misuse of public resources and misinformation. Both internal and external barriers and challenges should be identified and appropriate response developed [28]. How digital technologies influence the quality of discussions, whether actors engage constructively, whether dialogues are characterized by tolerance and resolving conflicts and disagreements, or whether they rather foster misinterpretation and increased polarization, should be further investigated [29]. It is also a matter of trust, where research for instance shows that face-to-face interactions are important in building trust and generate intention understanding in an international politics context [49], and that face-to-face negotiations have a higher level of initial trust between actors compared to online negotiations [50]. The use of technology at the different levels of online interaction (sharing, cooperation or collective action) has to be chosen deliberatively according to what is appropriate in that particular context, considering various risks. General challenges of digital divide, trust in the online context, cyber security and information overload have to be considered as well [21]. Additionally, the role of United Nations bodies in this context should be further researched, and elements that affect whether the desired effects of digitization are actualized clarified.

In the example of the Global Stocktake, a declaration will be adopted, and a question is how technology can be used to support the process of negotiating and adopting such a declaration (for instance with information provision to negotiators in the process). Another question relates to how technology could support work on implementation and follow up on decisions and commitments. A declaration that builds trust would have a clear statement of progress, commitments for climate action and means for implementation that responds to what is required to achieve the goal in the Paris Agreement.

3.4 Summary of the Research Agenda

In this paper we argue that while governance processes related to grand challenges such as the global climate crisis are moving online, this move currently seems experimental and largely lacking a fundamental understanding of the dynamics of online communities and online interactions. To address this problem, we discussed the example of the UN's Global Stocktake process in light of insights from research on online communities. Based on this discussion, we identified 10 questions across the three stages of online interactions (Table 2). We suggest that the questions constitute a research agenda to establish a necessary knowledge base for designing and implementing systems for online interactions in global governance contexts. Because this is a novel research area, the questions take an explorative approach of 'how' and 'what' questions, which can then be extended with other types of questions concerning 'when', 'where', and 'why' questions. The different levels of interaction will have different levels of complexity and associated design implications.

Table 2. Research agenda for online interaction in digital global governance

	Potential outcome	Suggested research questions	Global stocktake example
Sharing	*Improved basis for political decision making*	- How can technology facilitate collection of high-quality information from appropriate stakeholders? - How does technology influence the quality and flow of information? - How can technology be used to organize and disseminate information in comprehensible and inspiring ways to stakeholders?	*Collect, organize and disseminate climate reports*

(*continued*)

Table 2. (*continued*)

	Potential outcome	Suggested research questions	Global stocktake example
Cooperation	*Agreed upon knowledge status*	- What is the role of technology in synthesizing and leveraging actionable information? - What are the relationships between regulations, digital solutions and consensus forming? - What organizational, cultural, and competence-related frameworks are needed to facilitate cooperation that gains results?	*Synthesize national reports. Technical assessments leading to synthesis report*
Collective action	*Collective decisions binding for all individual members*	- What are the roles of technology in decision making processes? - What is the relationship between technology and trust in the negotiation process, and how to mitigate decisions resulting in "tragedy of the commons"? - What are the relationships between technology and co-production? - What is the role of technology in processes of accountability and follow up on decisions?	*Declaration*

4 Suggestions for Future Research

The core of our work is the identification of the research agenda for online interaction in digital global governance. While an in-depth discussion of all the potential theoretical approaches that may add value to address these questions is without the scope of our paper, we would like to propose some lenses that we find particularly interesting to address the research questions discussed above.

First, knowledge from the area of social movement organizations (SMO) could be of relevance for further studies within our context. SMO are collectives promoting social transformation through the mobilization of citizens for sustained political action [40]. Contrasting the more general concept of OC, SMO is focusing more directly on how online groups organize to achieve common objectives [51]. In particular, the research strands of SMO explore the role of collective actions, and the complex organizations needed to fulfill such goals. Future research addressing the need to understand how to organize to achieve collective actions (as proposed above) could be inspired by, for instance, the work of Mauss who almost fifty years back discussed the presence and connection between three main stakeholder groups within social movements; the outermost ring of a mass of sympathizers, the middle ring of a smaller number of active members committed to the movement's success, and the innermost ring of formal leaders and coordinators [52].

Mauss perspective is directly related to our next proposed theoretical lens. The stakeholder theory (ST) originated in management science in the 1980ies to improve organizations' capability to understand, predict and manage stakeholders (see e.g. Freeman [53, 54]). ST was later adapted to the eGovernment context (see e.g. Flak and Rose [55]) and has achieved considerable attention in this domain. We suggest that ST can be valuable in identifying and analyzing stakeholder complexity related to digital global governance. In particular, the theory of stakeholder identification and salience [48] may offer clarity on the salience of specific stakeholders or groups of stakeholders. Given the importance of transparency and legitimacy in digital global governance, we also argue that a recently proposed normative core of ST for the eGovernment context [56] can be used and further refined in this specific context.

Finally, we argue for the need to further investigate challenges related to the quality of the information in the context of online participation. Research on the use of OC for political participation [40, 41] show how some actors joined with the agenda of sabotaging the process, e.g., by posting false information within these forums. Hence, more research is needed to further investigate influence of technology on the distribution of misinformation (misleading or inaccurate information shared unconsciously), disinformation (false or misleading information shared intentionally) and fake news (false information packaged intentionally as real news) [57] in digital global governance.

4.1 Implications

The proposed research agenda will hopefully sensitize researchers of a critical knowledge gap that needs to be addressed with suggestions on how to embark on studies to reduce this gap. As such studies start to emerge, our initial research agenda should be critically assessed and developed further. Multi - and interdisciplinary research seem highly appropriate in this area as deep knowledge on governance of global phenomena needs to be matched with a deep understanding of digital technologies and the dynamics of online communities. Consequently, researchers can draw on a broad theory base in the quest to develop new knowledge in this area. We have suggested a few potentially valuable theoretical lenses in this paper.

The main audience for this paper is researchers with a potential interest in how digital technologies influence the governance of global phenomena. Nevertheless, we argue that the ideas and arguments in the paper also have practical relevance. Practitioners responsible for establishing and maintaining governance structures and processes to support the governance of global issues can benefit from being sensitized about the three stages of online interactions and the general dynamics of online communities. Moreover, the questions in our research agenda can also be applied from a more practical perspective to induce reflections on how different technologies may have different strengths and weaknesses depending on the stage they are being used in.

References

1. Hill, K.A., Hughes, J.E.: Cyberpolitics: Citizen Activism in the Age of the Internet. Rowman & Littlefield Publishers, Inc. (1998)
2. Ren, Y., Harper, F.M., Drenner, S., Terveen, L., Kiesler, S., Riedl, J., et al.: Building member attachment in online communities: applying theories of group identity and interpersonal bonds. MIS Q. **36**(3), 841–864 (2012)
3. Evans, D., Yen, D.C.: E-Government: evolving relationship of citizens and government, domestic, and international development. Gov. Inf. Q. **23**(2), 207–235 (2006)
4. Jreisat, J.: Governance in a globalizing world. Int. J. Public Adm. **27**(13–14), 1003–1029 (2004)
5. Moon, M.J., Welch, E.W., Wong, W.: What drives global E-governance? An exploratory study at a macro level. In: Proceedings of the 38th Hawaii International Conference on System Sciences (HICSS-38). Island of Hawaii (Big Island), p. 131 (1–10). Computer Societry Press (2005)
6. Qian, H.: Global perspectives on e-governance: from government-driven to citizen-centric public service delivery. In: 4th International Conference on Theory and Practice of Electronic Governance (ICEGOV 2010), Beijing, China, pp. 1–8. ACM (2010)
7. Zhao, F., Collier, A., Deng, H.: A multidimensional and integrative approach to study global digital divide and e-government development. Inf. Technol. People **27**(1), 38–62 (2014). https://doi.org/10.1108/Itp-01-2013-0022
8. Azad, B., Faraj, S., Goh, J.M., Feghali, T.: What shapes global diffusion of e-government: comparing the influence of national governance institutions. J. Glob. Inf. Manag. **18**(2), 85–104 (2010). https://doi.org/10.4018/jgim.2010040104
9. Tammpuu, P., Masso, A.: Transnational digital identity as an instrument for global digital citizenship: the case of Estonia's e-residency. Inf. Syst. Front. **21**(3), 621–634 (2019). https://doi.org/10.1007/s10796-019-09908-y
10. Navarra, D.D.: The architecture of global ICT programs: a case study of e-governance in Jordan. Inf. Technol. Dev. **16**(2), 128–140 (2010)
11. Frangonikolopoulos, C.A.: Global civil society and deliberation in the digital age. Int. J. Electron. Gov. **5**(1), 11–23 (2012)
12. UNFCCC: Conference of the Parties (COP) (2022). https://unfccc.int/process/bodies/supreme-bodies/conference-of-the-parties-cop. Accessed 16 Feb 2022
13. UNFCCC: May–June 2021 Climate Change Conference – sessions of the subsidiary bodies (2021). https://unfccc.int/event/may-june-2021-climate-change-conference-sessions-of-the-subsidiary-bodies. Accessed 02 Nov 2021
14. Klein, R., Harris, K., Bakhtaoui, I., Lager, F., Lindblom, A., Carson, M.: Building climate diplomacy back better: imagining the UNFCCC meetings of tomorrow. Stockholm Environment Institute, Stockholm (2021)
15. Ansell, C., Torfing, J.: Introduction: theories of governance. In: Ansell, C., Torfing, J. (eds.) Handbook on Theories of Governance. Edward Elgar Publishing Limited, pp. 1–17 (2016)

16. Zürn, M.: Global governance as multi-level governance. In: Handbook on Multi-Level Governance. Edward Elgar Smart Governance for Health and Well-Being: The Evidence, Cheltenham (2010)
17. Faraj, S., Jarvenpaa, S.L., Majchrzak, A.: Knowledge collaboration in online communities. Organ. Sci. **22**(5), 1224–1239 (2011)
18. Shirky, C.: Here Comes Everybody: The Power of Organizing Without Organizations. Penguin (2008)
19. Management SJo: Instructions for Research Notes and Book Reviews (2012). https://www.journals.elsevier.com/scandinavian-journal-of-management/policies/instructions-for-research-notes-and-book-reviews. Accessed 26 May 2022
20. Zürn, M.: Contested global governance. Glob. Policy **9**(1), 138–145 (2018)
21. Engvall, T., Flak, L.S.: Digital governance as a scientific concept. In: Charalabidis, Y., Flak, L.S., Pereira, G.V. (eds.) Scientific Foundations of Digital Governance and Transformation. PAIT, vol. 38, pp. 25–50. Springer, Cham (2022). https://doi.org/10.1007/978-3-030-92945-9_2
22. Bannister, F., Connolly, R.: New problems for old? Defining e-governance. In: Proceedings of the 44th Hawaii International Conference on System Sciences (HICSS-44), Kauai, Hawaii, USA, pp. 1–10. IEEE Computer Society Conference Publishing Services (2011)
23. European Commission: European Commission Digital Strategy. A digitally transformed, user-focused and data-driven Commission. European Commission, Brussels (2018)
24. European Commission: A European strategy for data, Brussels (2020)
25. Dawes, S.S.: Governance in the digital age: a research and action framework for an uncertain future. Gov. Inf. Q. **26**(2), 257–264 (2009). https://doi.org/10.1016/j.giq.2008.12.003
26. Hochtl, J., Parycek, P., Schollhammer, R.: Big data in the policy cycle: policy decision making in the digital era. J. Organ. Comput. Electron. Commer. **26**(1–2), 147–169 (2016)
27. Meijer, A.J., Löfgren, K.: The neglect of technology in theories of policy change. Int. J. Public Adm. Digit. Age **2**(1), 75–88 (2015). https://doi.org/10.4018/ijpada.2015010105
28. Edelmann, N.: Digitalisation and developing a participatory culture: participation, co-production, co-destruction. In: Charalabidis, Y., Flak, L.S., Pereira, G.V. (eds.) Scientific Foundations of Digital Governance and Transformation: Concepts, Approaches and Challenges, pp. 415–435. Springer, Cham (2022). https://doi.org/10.1007/978-3-030-92945-9_16
29. Tucker, J.A., Guess, A., Barberá, P., Vaccari, C., Siegel, A., Sanovich, S., et al.: Social media, political polarization, and political disinformation: a review of the scientific literature, 19 March 2018
30. Resca, A., Tozzi, M.L.: Offline and online communities: great differences and some similarities. In: Baskerville, R., De Marco, M., Spagnoletti, P. (eds.) Designing Organizational Systems, vol. 1, pp. 301–318. Springer, Heidelberg (2013). https://doi.org/10.1007/978-3-642-33371-2_16
31. Weber, M.: Economy and Society: An Outline of Interpretive Sociology. University of California Press (1978)
32. Preece, J.: Online communities: designing usability, supporting sociability. In: Industrial Management & Data Systems (2000)
33. Dellarocas, C.: Strategic manipulation of internet opinion forums: implications for consumers and firms. Manag. Sci. **52**(10), 1577–1593 (2006)
34. Leidner, D., Koch, H., Gonzalez, E.: Assimilating generation Y IT new hires into USAA's workforce: the role of an enterprise 2.0 system. MIS Q. Exec. **9**(4), 229–242 (2010)
35. Majchrzak, A., Faraj, S., Kane, G.C., Azad, B.: The contradictory influence of social media affordances on online communal knowledge sharing. J. Comput.-Mediat. Commun. **19**(1), 38–55 (2013)

36. Wasko, M., Faraj, S.: Why should I share? Examining social capital and knowledge contribution in electronic networks of practice. MIS Q. **29**(1), 35–57 (2005)
37. Butler, B., Sproull, L., Kiesler, S., Kraut, R.: Community effort in online groups: who does the work and why. In: Leadership at a Distance: Research in Technologically Supported Work, vol. 1, pp. 171–194 (2002)
38. Zammuto, R.F., Griffith, T.L., Majchrzak, A., Dougherty, D.J., Faraj, S.: Information technology and the changing fabric of organization. Organ. Sci. **18**(5), 749–762 (2007)
39. Fulk, J., DeSanctis, G.: Electronic communication and changing organizational forms. Organ. Sci. **6**(4), 337–349 (1995)
40. Braccini, A.M., Sæbø, Ø., Federici, T.: From the blogosphere into the parliament: the role of digital technologies in organizing social movements. Inf. Organ. **29**(3), 100250 (2019)
41. Sæbø, Ø., Federici, T., Braccini, A.M.: Combining social media affordances for organising collective action. Inf. Syst. J. **30**(4), 699–732 (2020)
42. Ostrom, E.: Governing the Commons: The Evolution of Institutions for Collective Action. Cambridge University Press, Cambridge (1990)
43. UN: Paris Agreement. In: Nations U (ed.) (2015)
44. UNFCCC: Global Stocktake (2022). https://unfccc.int/topics/global-stocktake
45. UNFCCC: About the Secretariat (2021). https://unfccc.int/about-us/about-the-secretariat. Accessed 02 Nov 2021
46. UNFCCC: Report of the Conference of the Parties serving as the meeting of the Parties to the Paris Agreement on the third part of its first session, held in Katowice from 2 to 15 December 2018 (2018)
47. Suh, K.-S., Lee, S., Suh, E.-K., Lee, H., Lee, J.: Online comment moderation policies for deliberative discussion–seed comments and identifiability. J. Assoc. Inf. Syst. **19**(3), 2 (2018)
48. Mitchell, R.K., Agle, B.R., Wood, D.J.: Toward a theory of stakeholder identification and salience: defining the principle of who and what really counts. Acad. Manag. Rev. **22**(4), 853–886 (1997)
49. Holmes, M., Wheeler, N.J.: Social bonding in diplomacy. Int. Theory **12**(1), 133–161 (2020)
50. Naquin, C.E., Paulson, G.D.: Online bargaining and interpersonal trust. J. Appl. Psychol. **88**(1), 113 (2003)
51. McAdam, D., Scott, W.R.: Organizations and movements. In: Social Movements and Organization Theory, p. 4 (2005)
52. Mauss, A.L.: Social Problems as Social Movements. Lippincott (1975)
53. Freeman, R.E.: Strategic Management: A Stakeholder Approach. Pitman, Boston (2010)
54. Lukas, J.: Ethik als Standard in der Beschaffung Werte und Normen als Gestaltungsausgangspunkt von Nicht-Regierungs-Organisationen. Springer, Wiesbaden (2017). https://doi.org/10.1007/978-3-658-17932-8
55. Flak, L.S., Rose, J.: Stakeholder governance: adapting stakeholder theory to e-government. Commun. Assoc. Inf. Syst. **16**, 642–664 (2005)
56. Rose, J., Flak, L.S., Sæbø, Ø.: Stakeholder theory for the E-government context: framing a value-oriented normative core. Gov. Inf. Q. **35**(3), 362–374 (2018). https://doi.org/10.1016/j.giq.2018.06.005
57. Kalsnes, B., Falasca, K., Kammer, A.: Scandinavian political journalism in a time of fake news and disinformation (2021)

Similarity-Based Dataset Recommendation Across Languages and Domains to Sentiment Analysis in the Electoral Domain

Jéssica Soares dos Santos(✉), Flavia Bernardini, and Aline Paes

Institute of Computing, Fluminense Federal University, Niterói, RJ, Brazil
jessicasoares@id.uff.br, {fcbernardini,alinepaes}@ic.uff.br

Abstract. Traditional machine learning classifiers usually fail at predicting labels for new data when their distribution differs from the training data distribution. This is particularly true with sentiment classifiers as the vocabulary and people's opinions rapidly evolve. Naturally, the problem aggravates when there are only a few or even none labeled instances in the target domain. In this paper, we propose a dataset recommendation method based on multilingual embeddings and similarity metrics to properly choose sentiment analysis datasets to be used as training set when labeled data is unavailable or scarce. We adopted the sentiment analysis of electoral domain as our case study, considering the complexity and difficulty for manually label millions of political social media opinions during the short period of campaigns. Our results suggest that dataset similarity may be considered, even when datasets belong to different languages, to minimize negative effects that may occur due to domain shift in sentiment classification tasks.

Keywords: Sentiment classification · Domain similarity · Multilingual sentiment classification · Dataset recommendation · Sentiment analysis · Electoral domain

1 Introduction

With the growth and popularity of social media platforms, data analysis using Web content has been pointed out as a new way to forecast outcomes in the real world and analyze opinion trends about several topics [9].

In recent years, approaches that use data from social media have been largely adopted to predict political election outcomes [18], most of them relying on sentiment analysis techniques [2]. Also called opinion mining, sentiment analysis is a research topic that focuses on extracting opinions and sentiments from natural language texts using computational methods.

Although there is a huge amount of electoral opinions that can be automatically gathered from social media in a fast way, adopting existing generic

R. Krimmer et al. (Eds.): ePart 2022, LNCS 13392, pp. 107–123, 2022.
https://doi.org/10.1007/978-3-031-23213-8_7

sentiment classifiers to label political data does not usually achieve good results [14]. So, in this work we chose to focus on the electoral scenario because it is a real world domain of large interest of the society that involves the difficulty to reliably label data, mainly due to the *high level of data complexity* and *time restriction* [17]. Regarding data complexity, this domain has a dynamic nature as it presents non-stationary data distributions, i.e., the vocabulary of terms used by people to express their opinions can change over time, mainly when critical sub-events (debates, speeches in interviews or political scandals) occur, causing degradation in the quality of predictions [4]. Additionally, such events can entail an irregular distribution of the data because during a debate or during the election day, there is a spike in information. Moreover, the dynamic characteristic of election domains [4] also entails that terms used in previous elections may not be the same used in the current election. The dynamic vocabulary that changes too fast according to election related sub events [4,15] may be one of the reasons to unsatisfactory electoral predictions. Regarding the time restriction, although there is a huge amount of electoral opinions available on the Web, due to the factors just pointed out, they need to be labeled during the short period of political campaigns [19]. Also, according to Lou (2020) [14], generic sentiment classification commercial systems usually do not perform well when applied to predict sentiment related to political data due to the complexity of this domain.

Dealing with none or limited labeled data in the domain of interest is a challenge that has been commonly tackled in sentiment analysis tasks. While there may be plenty of training labeled data in sentiment analysis domains, there is no guarantee that they follow the same distribution of the specific target domain of interest. The difference of the distribution between training and target datasets, called *domain shift*, may considerably impact the success rates of classification tasks on target data [7]. One factor that contributes to explain the decreasing of classification success rates when domain shift occurs is that expressions or words used to denote sentiment and characterize a sentence as positive, negative or neutral may vary from domain to domain [21]. For example, while for product reviews words such as *cheap* and *useful* are terms that denote positive sentiment, these words are not helpful for detecting positive sentences for other domains such as movie reviews or political domains. Analogously, while the word *unpredictable* denotes a positive sentiment for book or movie reviews, this word may indicate a negative orientation for automobile reviews (e.g.: "The steering of car is unpredictable") [10]. We elicited several reasons for sentiment prediction problems due to domain shift, as follows [1,7,13,23]:

- **Polysemy and Polarity Divergence:** Words can have different meanings and may denote different sentiments in different contexts;
- **Feature Divergence:** Terms that denote positive/negative sentiments may not be the same for source and target domains;
- **Sparsity:** Different datasets may have very different vocabularies and words may appear frequently in the target domain but may not appear (or appear rarely) in the source domain;
- **Writing Style:** Writing style may vary from domain to domain.

Reverse classification [7], sentiment graphs [20], prediction confidence, and language model perplexity [6] are some of the techniques that have been proposed in the literature to deal with domain shift and lack of annotated data in the target domain (see Sect. 2). However, those techniques are computationally intensive or require efforts for building graphs or training models. Also, these techniques were not designed to deal with multilingual comparisons.

In this context, we propose in this work a method for recommending training datasets based on measuring the semantic similarity between potential source datasets and the target dataset, considering data from different languages and domains. Our hypothesis is that datasets that contain similar content are more likely to lead to sentiment predictions closer to the ones obtained with classifiers trained with in-domain data, if they existed. The basic premise is that the target dataset has no labels and the similarity function must work in this scenario. The main advantage of our proposal is that there is no need to build models since training datasets are recommended based only on dataset similarity analysis.

To evaluate our proposal, we conducted an experimental analysis considering four target datasets from the electoral scenario domain, collected from Twitter, namely, one from the 2018 Brazilian presidential elections (Portuguese language); two from the 2012 and 2016 US presidential elections (English language); and one from the 2018 Colombian presidential elections (Spanish language). These real scenarios motivated our research due to (i) the lack of annotated data in this target domain and (ii) the difficulty to obtain these labels reliably due to time constraints.

Regarding the sixteen source datasets adopted in our experimental analysis, we consider (i) labeled datasets from different languages and different domains and (ii) datasets from different languages but in the same domain (elections scenario) to enlarge the possible recommendations. In the last case, we understand that due to the nature of the analyzed scenario – elections are events that occur regularly in democratic environments around the world –, datasets from the electoral domain may hold relevant shared information. In this way, we explore and compare semantic similarity across languages and domains to improve sentiment analysis in such a scenario. To that, examples are represented as embeddings, gathered from multilingual pre-trained models.

The evaluation methodology relies on the analysis of F1-score values achieved when using the recommended datasets for training classifiers. Our results indicate that dataset similarity may be considered to select proper sentiment analysis datasets. In short, electoral datasets were related to high values of semantic similarity, surpassing in some cases the results obtained with datasets in the same language. These results suggest that selecting training datasets from similar domains may contribute to better predictions regardless of language.

This paper is organized as follows: Sect. 2 presents a discussion about related work found in the literature. Section 3 presents our dataset recommendation method. Section 4 describes the use of this method applied to the electoral domain, and presents the evaluation methodology. Finally, Sect. 5 presents our conclusions and points out lines for future work.

2 Related Work

Many approaches have been proposed in the literature to deal with unlabeled data in the target domain. One example are the approaches for learning an embedding space to reduce the difference between source and target domains such as the Structural Correspondence Learning (SCL) [3,12]. A common representation is obtained using *pivot* features that occur frequently in both source and target domains to create a correspondence between features from these two datasets. Similarly, the work presented in [16] presents the Spectral Feature Alignment (SFA) that aligns domain-specific words from different domains into clusters, using domain independent words as a bridge.

Sentiment graphs (SG) are used by Wu and Huang (2016) [20] to extract polarity relations among words to compare different domains. Two types of relations are explored, namely: (i) sentiment coherent relation; and (ii) sentiment opposite relation. Those sentiment relations are identified according to manually selected rules. For example, words connected by conjunction prepositions are linked by sentiment coherent relation. On the other hand, words connected by adversative conjunctions are linked by sentiment opposite relations. A sentiment graph is created to each domain, where nodes represent words and edges represent the sentiment relations between words. A high domain similarity occurs when a pair of domains have many sentiment word pairs in common and the polarity relation scores of these word pairs are similar.

Reverse Classification Accuracy (RCA) methods are another possibility to choose proper source datasets, by estimating the performance drop of a model when evaluated on a new unlabeled target domain [7,8,24]. Basically, a given source dataset is used to build a classifier C_1 that will predict labels for the unlabeled target dataset. In turn, the new labeled dataset is used to train a classifier C_2. After that, the performance of classifier C_1 is compared to the performance of classifier C_2 using a subset of the source dataset as test data. The RCA* is a variation that uses one more classifier at the beginning of the process, to generate new labels for the source dataset and build from these new labels the classifier C_1. Confidence Based Measures (CBM) consists in using confidence scores – the certainty of the model over its predictions – as a similarity estimator. Therefore, this method requires not only the predicted labels but also the confidence scores [7].

Target Vocabulary Covered (TVC) was explored by Dai et el. (2019) [6] to choose proper source datasets based on similarity. It computes the number of words in the intersection between source and target datasets divided by the number of words of the target dataset. A Language Modeling (LM) based approach was also investigated by these authors. The basic idea is that, every time a language model trained on the source dataset finds a sentence in the target dataset that is very unlikely, then the model will assign a low probability and a high perplexity value. The closest source dataset is selected by taking into account all sentences from the target dataset. Finally, Dai et el. (2019) [6] explored an approach called Word Vector Variance (WVV) whose first step is to train a word vector on the source data using the skipgram model. Next, this trained word vector wv_1 is used to initialize weights of a new model trained on the target data, generating the trained word vector wv_2. The general idea is that the smaller the word vector variance, the more similar the two datasets are.

Table 1 presents a summary of the approaches aforementioned presented. Column *App.* indicates the name of the approach, column *Efforts* refers to the required efforts to implement the approach – where n = number of source datasets and m = number of target datasets, column *Sem.* indicates if the approach is able to capture semantic information (e.g.: using embeddings), column *Diff. lang.* indicates if the approach supports the selection of datasets from different languages, column *Based on* refers to the main elements used to improve/select source datasets.

Table 1. Approaches to deal with unlabeled target data

App.	Efforts	Sem.	Diff. lang.	Based on
SFA SCL	Learn a common feature representation meaningful for source and target domains	✓		Pivot features
SG	Create n sentiment graphs	✓		Sentiment graphs
TVC	Compute how many words are in the vocabularies intersection			Dataset vocabulary
WVV	Train $n + (n \times m)$ word vectors[a]	✓		Word vectors
LM	Train n language models	✓		Language models
RCA	Train $n + (n \times m)$ classifiers[a]	✓	✓	Classification models
RCA*	Train $2 \times n + (n \times m)$ classifiers[a]	✓	✓	Classification models
CB	Train n classifiers	✓	✓	Classification models
Our paper	Create similarity ranking	✓	✓	Similarity ranking

[a] If there is no intersection between the set of source and target datasets.

There are many aspects that differs our approach from related work. For example, RCA requires that two classifiers (or more as in the variation called RCA*) are built for each source dataset before computing similarity between source and target datasets. To measure dataset similarity between datasets using the LM approach, language models must be trained for each one of the source datasets. Also, the WVV requires the training of many word vectors. Techniques such as [3,12,16] may not work properly when source and training data do not share much information. The proposal of Wu and Huang (2016) [20] depends on manually selected rules to identify sentiment relations. We believe that rules are subjective and cannot be enough to represent similarity aspects in specific domains. Different from them, our method does not require any model or classifier to be trained. The dataset ranking is created based only on the analysis of semantic similarity between the source and target datasets. Then, our heuristic allows us to compare several datasets quickly and select the ones more similar to our domain of interest, being more appropriate for scenarios subjected to time restrictions and large amounts of unlabeled data, as in electoral scenarios. Furthermore, our proposed heuristic is independent of the selection of the algorithm that will be later adopted to train sentiment classifiers. The adoption of multilingual embeddings – which is another point that distinguishes our approach from the others – allows us to take advantage of data from different languages, a factor that may be very interesting in domains such as the electoral one.

3 Our Proposal

The diagram in Fig. 1 illustrates the process for executing our dataset recommendation method, whose activities (3.1 to 3.4) are described in what follows. The thin-bordered circle represents the beginning of the process. Process activities are denoted by round-cornered rectangles. The thick-bordered circle denotes the control that ends the activity. Dotted rectangles group related activities.

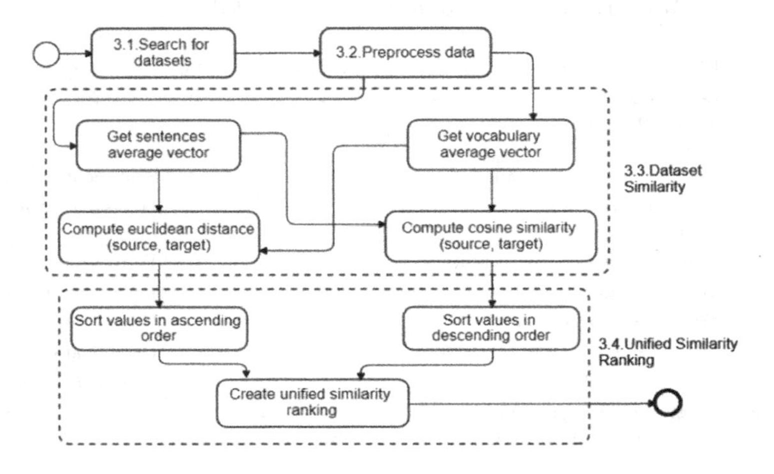

Fig. 1. Dataset recommendation process

3.1 Dataset Search

The first step of our method consists in acquiring labeled training data, assuming that existing knowledge may be useful for sentiment classification tasks when labeled data is not available in a given target domain. For this reason, we search for datasets publicly available in repositories such as Kaggle[1], GitHub[2], and Google Dataset Search[3]. To sum up, the requisites to the gathered datasets are that they must be sentiment analysis labeled datasets, given our focus on those problems.

3.2 Data Preprocessing

Preprocessing is applied to source and target datasets to remove noise data. Data cleaning includes: removing special characters, punctuation, accents and numbers;

[1] https://www.kaggle.com/.
[2] https://www.github.com/.
[3] https://datasetsearch.research.google.com/.

discard stop words (except the ones that denote contrast or negation); and convert all words to lower case. Also, sentiment classes of the datasets are analyzed maintaining only instances whose classes are common to all candidate datasets. Next, the datasets are converted to embeddings representations. Considering that datasets that belong to similar domains can be found in different languages, this approach uses the Universal Sentence Encoder Multilingual[4] [5,22], a pretrained model that was trained on 16 languages and is able to embed text from these different languages in a single vector space. The multilingual embeddings provided by this model allow us to compare and use datasets that belong to different languages.

3.3 Dataset Similarity

Our hypothesis is that selecting similar datasets before training machine learning classifiers is related to a greater chance to achieve satisfactory prediction results, when labeled data in the target dataset is not available. Two traditional similarity/distance metrics are used in this approach:

Cosine Similarity: Given two vectors $\mathbf{x}$ and $\mathbf{y}$, the cosine similarity measures the cosine of the angle between them, defined by Eq.(1).

$$s(\mathbf{x}, \mathbf{y}) = \frac{\mathbf{x} \cdot \mathbf{y}}{\sqrt{\mathbf{x} \cdot \mathbf{x}}\sqrt{\mathbf{y} \cdot \mathbf{y}}} \tag{1}$$

Euclidean Distance: Given two vectors $\mathbf{x}$ and $\mathbf{y}$, the euclidean distance measures the absolute value of the numerical difference of their coordinates, defined by Eq.(2).

$$d(x, y) = \sqrt{\sum_{i=1}^{n} (x_i - y_i)^2} \tag{2}$$

We are proposing to use four ways to analyze distributional semantic similarity between datasets:

Euclidean Distance Based on Sentences: For each sentence that belong to the dataset, we get the corresponding embedding according to the Universal Sentence Encoder Multilingual. After that, we compute a single embedding vector by averaging all of the sentence embeddings of the dataset. In this way, every dataset will have an average embedding that will represent its general context. Similarity between datasets is measured according to the Euclidean distance, i.e., the smaller the value of the euclidean distance between the average embeddings of a source-target dataset pair, the greater the similarity between this dataset pair.

Cosine Similarity Based on Sentences: This case is similar to the aforementioned case since an average embedding is computed for each dataset but the cosine similarity is used to measure dataset similarity. Therefore, the greater the value of cosine similarity between the average embeddings of a source-target dataset pair, the greater the similarity between this dataset pair.

[4] https://tfhub.dev/google/universal-sentence-encoder-multilingual/3.

Euclidean Distance Based on Vocabulary: In this case, we compare the vocabulary of two datasets to measure their similarity. We consider that the vocabulary of each dataset is composed by the words that appear in it and a single embedding vector is computed by averaging embeddings of these words. Each word is only considered once. Euclidean distance is used to compute the similarity between the dataset vocabularies.

Cosine Similarity Based on Vocabulary: Similarly to the last case mentioned, the vocabulary of the dataset pairs are compared using the cosine similarity metric.

These similarity methods will be used to compare pairs of datasets according to their content and to recommend training datasets.

3.4 Unified Similarity Ranking

Source datasets are sorted according to the similarity best values. All the similarity values for each source-target dataset pair are computed using four possibilities: *Euclidean Distance based on Sentences, Cosine Similarity based on Sentences, Euclidean Distance based on Vocabulary*, and *Cosine Similarity based on Vocabulary.* Therefore, four sorted lists are created to the given target dataset, where the most similar source datasets appear on the first positions and the most dissimilar appear on the last positions in each list. A unified ranking is created based on the $Hit(n)$ method, which computes how many times each source dataset appears in the top n positions in the four sorted lists, where n is an arbitrary number. Datasets that are likely to achieve good results are the ones associated with highest values in the unified ranking, whose cell values may be: 0 – source dataset does not appear within the first n positions in any of the lists ordered according to the values of similarity metrics; 1 – source dataset appears once in the first n positions; 2 – source dataset appears twice in the first n positions; 3 – source dataset appears three times in the first n positions; 4 – source dataset appears four times in the first n positions. In all the cases we are considering that they appear in the first positions of the lists ordered according to the values of similarity metrics.

4 Our Experimental Analysis

The dataset recommendation method[5] presented in Sect. 3 is applied to the electoral scenario domain. The evaluation methodology adopted to validate our method as well as our obtained results. It is important to mention that although the proposed dataset recommendation method is unsupervised, not requiring target dataset labels, we had to consider labeled target datasets in this experimental analysis to validate our results.

Datasets: We began by searching for existing labeled for the sentiment analysis task in publicly repositories to be used in our experiments, including Portuguese,

[5] https://github.com/sjessicasoaress/ds_recommender.

English, and Spanish languages. We selected Portuguese as it is our mother language, English because of the large available resources and the similarity of the 2016 USA electoral campaign and 2018 Brazilian campaign, and Spanish because it is a language with the same roots as Portuguese while vastly spoken in Latin America. We also searched for datasets that belong to the electoral domain to analyze this scenario as target. We found a total of 16 datasets, as follows:

1. *2018 BR Election PT* – tweets related to the 2018 Brazilian presidential elections;
2. *Restaurants PT* – opinions about Brazilian restaurants extracted from Foursquare;
3. *2016 US Election EN* – tweets about the 2016 United States presidential elections;
4. *GOP Debate EN*: tweets in english about a political debate that occurred in 2016;
5. *2012 US Election EN*: tweets about the 2012 United States presidential elections;
6. *TV PT*: tweets about Brazilian TV shows;
7. *Music Festival EN*: tweets about a US music festival;
8. *Urban Problems PT*: tweets about urban problems in Minas Gerais, a Brazilian state.
9. *Airlines EN*: tweets written in English about airlines;
10. *Movies 1 EN*: movie reviews in English;
11. *Movies PT*: movie reviews in Portuguese;
12. *Movies 2 EN*: movie reviews in English;
13. *Apple EN*: tweets written in English about Apple products;
14. *Airlines ES*: tweets about Spanish airlines;
15. *2018 CO Election ES*: tweets about the 2018 Colombian presidential elections;
16. *Sports ES*: tweets about sports in Spanish.

After selecting the datasets, we performed the preprocessing steps that removed special characters and stop words. Also, we had to discard instances associated with the class *neutral* since some of these datasets only have *positive* and *negative* instances.

Dataset Similarity Rankings: Semantic similarity was computed between each source-target dataset pair according to the similarity metrics presented in Sect. 3. While all the datasets are going to serve as source data, only the ones related to the electoral domain are selected as target, since this is our focus in this paper. Tables 2, 3, 4, and 5, present the results for the *Euclidean Distance based on Sentences*, *Cosine Distance based on Sentences*, *Euclidean Distance based on Vocabulary* and *Cosine Distance based on Vocabulary* metrics, respectively. The columns of these tables represent target datasets and rows represent source datasets. The best similarity values are highlighted with stronger colors, while the worst are highlighted with lighter colors.

The unified similarity ranking was built according to the $Hit(n)$ method adopting $n = 5$. Therefore, each table cell value corresponds to the number

Table 2. Euclidean distance based on sentences

	2018 BR Election PT	2016 US Election EN	2012 US Election EN	2018 CO Election ES
2018 BR Election PT	-	0.5227	0.5254	0.4133
Restaurants PT	0.6374	0.6332	0.6417	0.4945
2016 US Election EN	0.5227	-	0.4517	0.4208
GOP Debate EN	0.5309	0.4568	0.4254	0.4910
2012 US Election EN	0.5254	0.4517	-	0.4384
TV PT	0.5648	0.5892	0.5644	0.4083
Music Festival EN	0.8554	0.8170	0.7911	0.7268
Urban Problems PT	0.5756	0.5811	0.5647	0.4842
Airlines EN	0.6392	0.5633	0.5532	0.4597
Movies 1 EN	0.6223	0.5467	0.5447	0.4408
Movies PT	0.6386	0.6062	0.5867	0.4787
Movies 2 EN	0.6695	0.6098	0.5893	0.5133
Apple EN	0.6039	0.5276	0.5093	0.4357
Airlines ES	0.5897	0.5810	0.5819	0.4252
2018 CO Election ES	0.4133	0.4208	0.4384	-
Sports ES	0.5897	0.5789	0.5643	0.3925

Table 3. Cosine similarity based on sentences

	2018 BR Election PT	2016 US Election EN	2012 US Election EN	2018 CO Election ES
2018 BR Election PT	-	0.5677	0.5651	0.7107
Restaurants PT	0.2780	0.1911	0.1747	0.3527
2016 US Election EN	0.5677	-	0.6415	0.6267
GOP Debate EN	0.5899	0.6692	0.7148	0.5598
2012 US Election EN	0.5651	0.6415	-	0.5957
TV PT	0.4351	0.2937	0.3576	0.5536
Music Festival EN	0.1527	0.1715	0.2291	0.2607
Urban Problems PT	0.4670	0.3926	0.4299	0.4792
Airlines EN	0.2253	0.3147	0.3456	0.3764
Movies 1 EN	0.2397	0.3283	0.3394	0.3868
Movies PT	0.2569	0.2377	0.2923	0.3672
Movies 2 EN	0.1984	0.2485	0.3039	0.2990
Apple EN	0.2829	0.3718	0.4239	0.3917
Airlines ES	0.3533	0.2747	0.2780	0.4723
2018 CO Election ES	0.7107	0.6267	0.5957	-
Sports ES	0.3901	0.3306	0.3689	0.6001

Table 4. Euclidean distance based on vocabulary

	2018 BR Election PT	2016 US Election EN	2012 US Election EN	2018 CO Election ES
2018 BR Election PT	-	0.0831	0.0941	0.0555
Restaurants PT	0.1168	0.1259	0.1092	0.1124
2016 US Election EN	0.0831	-	0.0484	0.0912
GOP Debate EN	0.1078	0.0587	0.0504	0.1024
2012 US Election EN	0.0941	0.0484	-	0.0932
TV PT	0.0858	0.1047	0.0856	0.0848
Music Festival EN	0.1541	0.1208	0.0857	0.1453
Urban Problems PT	0.0906	0.0963	0.0978	0.0906
Airlines EN	0.1216	0.0802	0.0674	0.1152
Movies 1 EN	0.1115	0.0730	0.0830	0.1091
Movies PT	0.0669	0.0686	0.0735	0.0706
Movies 2 EN	0.1008	0.0654	0.0724	0.1066
Apple EN	0.1527	0.1136	0.0935	0.1428
Airlines ES	0.0970	0.1085	0.0932	0.0814
2018 CO Election ES	0.0555	0.0912	0.0932	-
Sports ES	0.0680	0.0881	0.0797	0.0595

Table 5. Cosine similarity based on vocabulary

	2018 BR Election PT	2016 US Election EN	2012 US Election EN	2018 CO Election ES
2018 BR Election PT	-	0.9905	0.9878	0.9961
Restaurants PT	0.9812	0.9783	0.9837	0.9829
2016 US Election EN	0.9905	-	0.9968	0.9887
GOP Debate EN	0.9838	0.9952	0.9965	0.9857
2012 US Election EN	0.9878	0.9968	-	0.9882
TV PT	0.9901	0.9851	0.9901	0.9903
Music Festival EN	0.9677	0.9803	0.9901	0.9716
Urban Problems PT	0.9885	0.9872	0.9868	0.9889
Airlines EN	0.9791	0.9911	0.9938	0.9819
Movies 1 EN	0.9829	0.9927	0.9906	0.9839
Movies PT	0.9938	0.9935	0.9926	0.9933
Movies 2 EN	0.9859	0.9941	0.9928	0.9846
Apple EN	0.9673	0.9822	0.9880	0.9722
Airlines ES	0.9868	0.9837	0.9880	0.9910
2018 CO Election ES	0.9961	0.9887	0.9882	-
Sports ES	0.9940	0.9895	0.9914	0.9952

Table 6. Unified similarity ranking summary

	2018 BR Election PT	2016 US Election EN	2012 US Election EN	2018 CO Election ES
2018 BR Election PT	-	2	2	4
Restaurants PT	0	0	0	0
2016 US Election EN	4	-	4	2
GOP Debate EN	2	4	4	1
2012 US Election EN	2	4	-	1
TV PT	3	0	0	3
Music Festival EN	0	0	0	0
Urban Problems PT	1	1	1	0
Airlines EN	0	0	2	0
Movies 1 EN	0	2	0	0
Movies PT	2	2	2	2
Movies 2 EN	0	2	2	0
Apple EN	0	1	1	0
Airlines ES	0	0	0	3
2018 CO Election ES	4	2	2	-
Sports ES	2	0	0	4

of hits, i.e., how many times a given source-target dataset is at top-five first positions according to the similarity results presented in Tables 2, 3, 4 and 5. Table 6 illustrates the similarity ranking as a heat map.

For each column in Table 6 we can see the most recommended source datasets for a given target dataset as being the ones with stronger colors. Columns represent target datasets and rows represent source datasets.

Validation of Our Similarity Heuristics for Recommending Datasets: We tested all pairs of source-target datasets with a set of five traditional machine learning algorithms that are widely adopted to deal with natural language processing tasks, using the same pre-trained embeddings employed during the similarity analysis: Support Vector Machine (SVM), Logistic Regression (LR), Decision Tree (DT), Multi-Layer Perceptron (MLP), and XGBoost. From these sixteen datasets, four belong to the target domain of this experimental analysis

(2018 BR Election PT, 2016 US Election EN, 2012 US Election EN, 2018 CO Election ES). For each of the five machine learning algorithms, we trained different classifiers using each one of the source datasets individually as training data. The classifiers were applied on the four electoral datasets, individually. Table 7 presents the F1-score results of each classifier. Due to limitation space, datasets are identified by a number instead of its name. The numeric identifier is the same as the one used before: 1 – corresponds to the 2018 BR Election PT, 3 – corresponds to the 2016 US Election EN, 5 – corresponds to the 2012 US Election EN, and 15 – corresponds to the 2018 CO Election ES. Table 7 columns refer to the identifier of the source dataset used to train classifiers. Rows refer to the identifier of the target dataset followed by the abbreviation of the machine learning algorithm. The best F1-score values are highlighted with stronger colors, while the worst are highlighted with lighter colors. For each combination of target dataset–algorithm, the obtained F1-score classification results – when using the different source datasets as training data – are stored in a list which is sorted according to F1-score best values. From that, the $Hit(n)$ method is used to create a unified F1-score ranking for each target dataset. In this ranking, each source dataset is associated with the number of times it appeared in the top 5 first positions considering the sorted lists of F1-score values. Table 8 illustrates the unified F1-score ranking as a heat map. Cells present how many times a given source dataset is at top 5 first positions for each target dataset according to the F1-score results presented in Table 7.

Table 7. F1-score summary

	1	2	3	4	5	6	7	8	9	10	11	12	13	14	15	16
1 - SVM	-	0.72	0.72	0.58	0.72	0.60	0.63	0.46	0.75	0.47	0.56	0.63	0.70	0.72	0.73	0.78
1 - LR	-	0.72	0.73	0.57	0.74	0.61	0.59	0.46	0.75	0.44	0.57	0.61	0.73	0.76	0.72	0.76
1 - DT	-	0.52	0.61	0.54	0.57	0.62	0.5	0.48	0.58	0.50	0.56	0.54	0.57	0.65	0.62	0.63
1 - MLP	-	0.69	0.66	0.51	0.71	0.62	0.48	0.47	0.66	0.48	0.61	0.63	0.65	0.70	0.67	0.77
1 - XGBoost	-	0.73	0.75	0.67	0.72	0.67	0.57	0.5	0.72	0.54	0.66	0.63	0.66	0.72	0.74	0.72
3 - SVM	0.66	0.53	-	0.63	0.68	0.59	0.6	0.52	0.60	0.54	0.63	0.63	0.65	0.61	0.58	0.63
3 - LR	0.66	0.53	-	0.63	0.69	0.55	0.56	0.53	0.62	0.55	0.64	0.63	0.65	0.60	0.60	0.61
3 - DT	0.55	0.5	-	0.55	0.56	0.53	0.51	0.49	0.54	0.51	0.51	0.52	0.54	0.54	0.53	0.56
3 - MLP	0.63	0.49	-	0.61	0.68	0.51	0.57	0.54	0.61	0.51	0.55	0.52	0.63	0.57	0.58	0.63
3 - XGBoost	0.62	0.49	-	0.62	0.69	0.53	0.58	0.54	0.60	0.54	0.62	0.6	0.59	0.61	0.62	0.60
5 - SVM	0.64	0.58	0.71	0.55	-	0.65	0.58	0.57	0.67	0.58	0.58	0.57	0.65	0.61	0.55	0.68
5 - LR	0.63	0.60	0.70	0.51	-	0.6	0.57	0.56	0.67	0.53	0.59	0.57	0.68	0.63	0.55	0.68
5 - DT	0.59	0.49	0.62	0.52	-	0.52	0.54	0.53	0.52	0.47	0.52	0.53	0.6	0.55	0.57	0.54
5 - MLP	0.6	0.54	0.66	0.55	-	0.61	0.53	0.59	0.63	0.54	0.55	0.59	0.63	0.58	0.56	0.63
5 - XGBoost	0.61	0.55	0.69	0.54	-	0.56	0.57	0.61	0.62	0.56	0.59	0.63	0.60	0.58	0.58	0.65
15 - SVM	0.67	0.62	0.61	0.56	0.61	0.65	0.64	0.57	0.67	0.53	0.59	0.60	0.65	0.63	-	0.66
15 - LR	0.67	0.63	0.63	0.57	0.62	0.65	0.63	0.55	0.68	0.53	0.60	0.61	0.66	0.65	-	0.67
15 - DT	0.59	0.54	0.59	0.52	0.55	0.58	0.51	0.53	0.60	0.49	0.53	0.53	0.56	0.57	-	0.58
15 - MLP	0.65	0.59	0.62	0.51	0.58	0.63	0.60	0.57	0.63	0.50	0.55	0.57	0.63	0.62	-	0.66
15 - XGBoost	0.67	0.62	0.66	0.62	0.64	0.65	0.63	0.56	0.68	0.53	0.60	0.59	0.65	0.65	-	0.66

Table 8. Unified F1-score ranking summary

	2018 BR Election PT	2016 US Election EN	2012 US Election EN	2018 CO Election ES
2018 BR Election PT	-	5	3	5
Restaurants PT	3	0	0	0
2016 US Election EN	4	-	5	2
GOP Debate EN	0	5	0	0
2012 US Election EN	3	5	-	0
TV PT	1	0	2	5
Music Festival EN	0	0	0	0
Urban Problems PT	0	0	0	0
Airlines EN	3	1	4	5
Movies 1 EN	0	0	0	0
Movies PT	0	3	0	0
Movies 2 EN	0	2	1	0
Apple EN	0	3	4	3
Airlines ES	3	0	1	0
2018 CO Election ES	4	1	1	-
Sports ES	4	2	4	5

Discussion: We compared the unified similarity ranking to the unified F1-score ranking to verify if the recommended datasets are among the ones associated with higher values in the unified similarity ranking. By looking at column *2018 BR Election PT* in Table 6, two datasets are recommended in first place (number of hits = 4), namely: *2016 US Election EN* and *2018 CO Election ES*. Both datasets also appeared in the first position in the F1-score ranking for the column *2018 BR Election PT* in Table 8 (number of hits = 4).

Observing results for column *2016 US Election EN* in Table 6, two datasets are recommended in first place (number of hits = 4), namely: *2012 US Election EN* and *GOP Debate EN*. Both datasets also appeared in the first position in the F1-score ranking for the column *2016 US Election EN* in Table 8 (number of hits = 5). For column *2012 US Election EN* in Table 6, two datasets are recommended in first place (number of hits = 4), namely: *2016 US Election EN* and *GOP Debate EN*. While dataset *2016 US Election EN* also appeared in the first position in the F1-score ranking for the column *2012 US Election EN* in Table 8 (number of hits = 5), *GOP Debate EN* did not appear among the top 5 first positions. We believe that this may occurr when the datasets have similar content but there is a high level of polarity divergence between them. This would be a limitation of our method, although in practice it is not possible to measure polarity divergence in cases where there is no data annotated in the target domain as target labels are not available. The results for the target dataset *2018 CO Election ES* show two source datasets being recommended in the first position (number of hits = 4), namely, *2018 BR Election PT* and *Sports ES* datasets. Both datasets also appeared in the first position in the F1-score ranking for the column *2018 CO Election ES* in Table 8 (number of hits = 5).

Finally, we compared our method to other methods for choosing proper training datasets presented in Sect. 2. From the approaches mentioned in Sect. 2, we were not able to compare our method to the SG strategy as it requires sentiment graphs to each one of the source datasets. Also, we do not consider the SFA/SCL strategies in our comparison as they are not based on source dataset selection. Instead, they focus on creating a mapping between source and the target domain, not considering the existence of multiple source datasets. We believe that this kind of strategy may be useful in cases where only a single source dataset is available to the target task. The results of our comparison are presented in Table 9 and details are available online[6]. All the values presented in Table 9 are the identifiers of the datasets that were recommended by each one of the dataset selection methods. The first column refers to the identifier of the target datasets (1-2018 BR Election PT, 3-2016 US Election EN, 5-2012 US Election EN, 15-2018 CO Election ES). The column *F1 ranking* refers to the datasets that are associated with the highest values in the F1 unified ranking and, therefore, are the ones that achieved the highest F1-score in our experiments. The remainder columns refer to the approaches adopted (TVC, WVV, RCA, RCA*, LM, CB) for selecting training datasets. As we can notice, the strategy presented in this paper achieved better recommendations than the other ones. This is probably because our case study involved datasets from different languages and many of these strategies were not designed to deal with multilingual dataset comparison (e.g.: WVV and TVC). Another factor that may explain this is that some approaches depend on very large source datasets (e.g.: LM) to better identify semantic relationships.

Table 9. Recommended datasets by approach

	TVC	WVV	RCA	RCA*	LM	CB	Our method	F1 ranking
1	11	6	4, 8	3	6	9	3, 15	3, 15, 16
3	12	6	8	4	4	9	4, 5	1, 4, 5
5	12	6	4	3	4	1	3, 4	3
15	16	11	8	10	16	16	1, 16	1, 6, 9, 16

5 Conclusions

The electoral scenario is a real world domain of interest to the society where it is difficult to label examples reliably. In this paper, we presented a dataset recommendation method for helping one to choose, from a set of sentiment analysis labeled datasets, possible candidates to induce classifiers to unlabeled data. The recommendation method relies on analyzing dataset semantic similarity between labeled and unlabeled examples. The usage of multilingual embeddings as the vectorization technique allows us to compare and use datasets from different languages (English, Portuguese, Spanish) as classifiers training data. We suggest to use two similarity measures, namely, cosine similarity and euclidean distance combined with building vectors from averaging examples and averaging words vectors. The main advantage of our method regarding other approaches in the literature to select proper training datasets is that only similarity metrics between

[6] https://bit.ly/3FC3qC5.

the datasets are calculated, being much less expensive than building models or classifiers for all the possible training datasets. Also, we are able to consider source datasets from different languages as we explore multilingual embeddings.

To evaluate the quality of the dataset recommendation, we analyze both dataset similarity measures and F1-score ranking results. Our findings suggest that the analysis of dataset semantic similarity can be beneficial when one needs to choose a dataset to be used as starting point for training classifiers. The presented investigation also shows that results related to high values of semantic similarity between datasets could in some cases surpass results obtained with datasets in the same language, leading us to believe that similar domains can contribute to better results regardless of language. This specific point can be observed by looking at the prediction results obtained using electoral datasets as training data. Some of the source datasets that appeared out of the top five first positions in our ranking were also able to achieve good predictions results. This factor may have occurred due to the existence of other common characteristics between the source-target dataset pairs that were not captured by using only the similarity metrics considered in this paper. Also, factors such as unbiased data can improve the generalization ability of a model, what could explain models performing well despite dataset similarity [11]. It is important to emphasize that sometimes heuristic inference methods may not be able to present the optimal datasets but being able to point out datasets that can achieve values that are close to the optimal ones. However, adopting dataset recommendation heuristics to select similar datasets beforehand has the advantage of not having to run all possible experiments to get a satisfactory result, saving time and reducing computational costs. Furthermore, if there is no label for the target domain, it is not possible to be sure which is the most proper source dataset and strategies like the one proposed in this paper can be adopted. As a future work, we intend to investigate the use of other similarity metrics to improve the recommendations of our method. Also, we intend to test if the combination of subsets of the recommended datasets may improve sentiment classification.

Acknowledgement. This research was supported by the Brazilian Research CNPq APQ Universal (Grant 421608/2018-8), CNPq Research Grant 311275/2020-6, FAPERJ Research grant E26/202.914/2019 (247109), Microsoft Research Grant and Coordenação de Aperfeiçoamento de Pessoal de Nível Superior - Brazil (CAPES).

References

1. Al-Moslmi, T., Omar, N., Abdullah, S., Albared, M.: Approaches to cross-domain sentiment analysis: A systematic literature review. IEEE Access **5**, 16173–16192 (2017)
2. Bilal, M., Gani, A., Marjani, M., Malik, N.: Predicting elections: social media data and techniques. In: 2019 International Conference on Engineering and Emerging Technologies (ICEET), pp. 1–6. IEEE (2019)
3. Blitzer, J., McDonald, R., Pereira, F.: Domain adaptation with structural correspondence learning. In: Proceedings of the 2006 Conference on Empirical Methods in Natural Language Processing, pp. 120–128 (2006)

4. Calais Guerra, P.H., Veloso, A., Meira Jr, W., Almeida, V.: From bias to opinion: a transfer-learning approach to real-time sentiment analysis. In: Proceedings of the 17th ACM SIGKDD International Conference on Knowledge Discovery and Data Mining, pp. 150–158 (2011)
5. Chidambaram, M., et al.: Learning cross-lingual sentence representations via a multi-task dual-encoder model. arXiv preprint arXiv:1810.12836 (2018)
6. Dai, X., Karimi, S., Hachey, B., Paris, C.: Using similarity measures to select pretraining data for NER. In: Proceedings of the 2019 Conference of the North American Chapter of the Association for Computational Linguistics: Human Language Technologies, Volume 1 (Long and Short Papers), pp. 1460–1470 (2019)
7. Elsahar, H., Gallé, M.: To annotate or not? Predicting performance drop under domain shift. In: Proceedings of the 2019 Conference on Empirical Methods in Natural Language Processing and the 9th International Joint Conference on Natural Language Processing (EMNLP-IJCNLP). pp. 2163–2173 (2019)
8. Fan, W., Davidson, I.: Reverse testing: an efficient framework to select amongst classifiers under sample selection bias. In: Proceedings of the 12th ACM SIGKDD International Conference on Knowledge Discovery and Data Mining, pp. 147–156 (2006)
9. Ghani, N.A., Hamid, S., Hashem, I.A.T., Ahmed, E.: Social media big data analytics: a survey. Comput. Hum. Behav. **101**, 417–428 (2019)
10. Joshi, M., Prajapati, P., Shaikh, A., Vala, V.: A survey on sentiment analysis. Int. J. Comput. Appl. **163**(6), 34–38 (2017)
11. Kouw, W.M., Loog, M.: An introduction to domain adaptation and transfer learning. arXiv preprint arXiv:1812.11806 (2018)
12. Li, N., Zhai, S., Zhang, Z., Lou, B.: Structural correspondence learning for cross-lingual sentiment classification with one-to-many mappings. In: Proceedings of the AAAI Conference on Artificial Intelligence, vol. 31 (2017)
13. Li, Y., Guo, H., Zhang, Q., Gu, M., Yang, J.: Imbalanced text sentiment classification using universal and domain-specific knowledge. Knowl. Based Syst. **160**, 1–15 (2018)
14. Lou, B.: Sentiment Analysis: Mining Opinions, Sentiments, and Emotions. Studies in Natural Language Processing, 2 edn. Cambridge University Press, (2020). https://doi.org/10.1017/9781108639286
15. Mahendiran, A., et al.: Discovering evolving political vocabulary in social media. In: 2014 International Conference on Behavioral, Economic, and Socio-Cultural Computing (BESC2014), pp. 1–7. IEEE (2014)
16. Pan, S.J., Ni, X., Sun, J.T., Yang, Q., Chen, Z.: Cross-domain sentiment classification via spectral feature alignment. In: Proceedings of the 19th International Conference on World Wide Web, pp. 751–760 (2010)
17. Santos, J.S., Bernardini, F., Paes, A.: Measuring the degree of divergence when labeling tweets in the electoral scenario. In: Anais do X Brazilian Workshop on Social Network Analysis and Mining. pp. 127–138. SBC (2021)
18. Santos, J.S., Bernardini, F., Paes, A.: A survey on the use of data and opinion mining in social media to political electoral outcomes prediction. Soc. Netw. Anal. Min. **11**(1), 1–39 (2021)
19. Santos, J.S., Paes, A., Bernardini, F.: Combining labeled datasets for sentiment analysis from different domains based on dataset similarity to predict electors sentiment. In: Proceedings of the 8th Brazilian Conference on Intelligent Systems (BRACIS), pp. 455–460. IEEE (2019)

20. Wu, F., Huang, Y.: Sentiment domain adaptation with multiple sources. In: Proceedings of the 54th Annual Meeting of the Association for Computational Linguistics (Volume 1: Long Papers), pp. 301–310 (2016)
21. Wu, F., Huang, Y., Yuan, Z.: Domain-specific sentiment classification via fusing sentiment knowledge from multiple sources. Inf. Fus. **35**, 26–37 (2017)
22. Yang, Y., et al.: Multilingual universal sentence encoder for semantic retrieval. arXiv preprint arXiv:1907.04307 (2019)
23. Zhang, Y., Hu, X., Li, P., Li, L., Wu, X.: Cross-domain sentiment classification-feature divergence, polarity divergence or both? Pattern Recogn. Lett. **65**, 44–50 (2015)
24. Zhong, E., Fan, W., Yang, Q., Verscheure, O., Ren, J.: Cross validation framework to choose amongst models and datasets for transfer learning. In: Balcázar, J.L., Bonchi, F., Gionis, A., Sebag, M. (eds.) ECML PKDD 2010. LNCS (LNAI), vol. 6323, pp. 547–562. Springer, Heidelberg (2010). https://doi.org/10.1007/978-3-642-15939-8_35

Genres of Participation in Social Networking Systems: A Study of the 2021 Norwegian Parliamentary Election

Marius Rohde Johannessen(✉)

School of Business, University of South-Eastern Norway, Po Box 4, 3199 Borre, Norway
marius.johannessen@usn.no

Abstract. eParticipation covers a range of activities, from online voting to deliberation. In this paper, I examine the activity called e-campaigning, political parties use of digital channels during election campaigns. Norwegian political parties have used digital channels for campaigning since 2001, and social media since 2009. In this study, I present the findings from the Norwegian parliamentary election campaign of 2021. Using genre theory, I examine how Norwegian political campaigning in social media has evolved between 2009 and the latest election in 2021. The findings indicate that social media use has stabilized between the 2017 and 2021 elections, with communication genres established in 2017 being similar to the ones found in 2021. Finally, I discuss how the way social media is currently being used reflects the constitutional goal of including citizens in political participation.

Keywords: eParticipation · Social media · Online campaigning · Social networking systems · Genre theory · Norway

1 Introduction

Participation, whether digital or offline, is not one thing. Rather, participation and eParticipation describe a wide range of different activities, from voting in elections via campaigning to dialogue between citizens and politicians or between different groups of citizens [1, 2]. Dialogue, or deliberation, where people meet and freely discuss political matters is arguably the "highest" form of participation [3], along with actual power-sharing between politicians and citizens [4]. Democratic theory varies in how participation is understood, and what part participation should play in democracy. Ferree and colleagues [5] outline four different models; Representative Liberal, Participatory Liberal, Discursive, and Constructionist. These models refer to different normative criteria for democracy in terms of who should speak, the content of the process (what), style of speech preferred (how), and the relationship between discourse and decision-making (outcomes) that is sought (or feared) [p. 290].

In the Norwegian constitution, article 100 states that "the authorities of the state shall create conditions that facilitate open and enlightened public discourse" [6]. Norwegian

Published by Springer Nature Switzerland AG 2022
R. Krimmer et al. (Eds.): ePart 2022, LNCS 13392, pp. 124–136, 2022.
https://doi.org/10.1007/978-3-031-23213-8_8

democracy is thus closer to what Ferree and colleagues call the participatory liberal model, where democracy is rooted in representative traditions, but citizen participation and dialogue is wanted also between elections and public debate is seen as a strength for democracy.

Despite this focus on participation, the Norwegian public sphere is not immune to the recent trends of fake news, polarization, and disinformation. Several recent theses have found examples of disinformation in relation to the covid-19 pandemic, aggressive echo chambers for "male culture" as well as polarization and distrust related to far-right groups in social media [7–10]. In this paper, I apply genre theory to examine the Norwegian social media election campaign of 2021. While campaigning is a distinct form of participation [2, 11], the emphasis on dialogue in Norwegian democracy can be said to provide normative guidelines for communication in the election campaign. I also draw on data from my previous studies of the campaigns in 2009, 2013 and 2017 to examine if and how social media campaigning in Norway has changed over the duration of four election campaigns.

This paper thus responds to the call for more research on the current changes to the public sphere [12] by examining communication genres in the 2021 Norwegian parliamentary election. By following the same research design as I did for previous elections, I seek to answer the following research question: *Which genres were used during the 2021 election campaign, and how have they evolved compared to previous elections?*

2 Related Research

2.1 Theoretical Lens: The Public Sphere - Participation Through Campaigning

A cynic would perhaps see political campaigning as a form of marketing, as it is all about getting the message out to the public, telling people about all the good parts of the political party's program. The election campaign has a very big influence on the outcome of the parliamentary election. More than 40% of Norwegian voters wait until the final weeks of the campaign before deciding who gets their vote, and many change their mind several times during the campaign, and since the sixties the trend has increasingly moved towards not voting for the same party twice [13]. Younger voters are more likely to cast their vote differently from one election to the next. Historically, Norwegian newspapers belonged to one of the two major political parties (Labour and the Conservatives). When the Norwegian newspapers became politically independent, political parties lost the power to decide what should be on the public agenda [14]. Since then, the media has taken over the agenda-setting role, and are trying to write about the things they believe voters are concerned about [13, 14]. Taking back control was one of the reasons why Norwegian political parties started using the Internet for campaigning in 2001 [15], and in 2007 first began experimenting with campaigning in social media [16]. In the 2017 election, traditional media such as TV news and newspapers were still the most frequently used media, but social media (Facebook) was for the first time on the list. In the last week of the election, 48% reported watching TV news daily, 26% read the online newspaper VG and 23% visited Facebook daily searching for political content [13].

While campaigning is a distinct form of participation [2, 11], the emphasis on dialogue in Norwegian democracy can be said to provide normative guidelines for communication in the election campaign. Thus, we can evaluate the election campaign in terms of liberal participatory democracy [5] and the deliberative ideals of the public sphere, where participation should be rational, relevant, polite and allow everyone to speak [17]. To apply the public sphere in the current fragmented media landscape means we should discuss this not as one, but as several sometimes-overlapping spheres [18], which can be both representing the majority, counter-cultures or a mix of both [19]. Further, today's public sphere has moved us away from being passive spectators of a media-controlled public debate. We have become active participants on social media, but at the same time more pessimistic voices claim we have seen the public sphere move from rational discourse to "noise. Senseless memes, "one-liners", half-truths (and many lies), vulgarity, insults, and of course, cheap partisan propaganda" [20].

2.2 Analytical Lens: Genre Theory

A genre can be defined as "a conventional category of discourse based in large-scale typification of rhetorical action" [21]. Genre theory can be applied to classify communication practices, and has been a valuable tool for studying online democracy [22], as well as for modelling potential democracy systems [23]. Genre theory provides us with a lens for detailed understanding of political communication [24]. Genres are recognized by having similar form and content, where form refers to physical and linguistic features, and content to themes and topics of the genre [25], and with digital communication also the functionality of the medium [26], as the affordances of the medium influence the ways in which actors engage in meaning-making [27]. Genres can be defined by examining form, functionality and content, by using the 5w1h-method [28]: *Where* tells us where the communication takes. *Why* explains the purpose of the genre. *When* refers to the time where communication takes place. *Who* defines the actors involved in communication, the sender and receiver of the genre. *What* is the content of the genre and *How* describes the technical needs for delivery of the genre. The genres used by a given community can be seen as a genre system [29], and this system can reveal the communicative practices shaped by community members in response to norms, events, time pressure and media capabilities [24]. By studying communication genres and communicative practices instead of the technology used to communicate, we can discover how communication changes and evolves over time [24], and by including the technological functionality of the medium the genre is enacted within, we can better understand the interplay between the social and the technical [26], allowing for a deeper understanding of communication situated in a specific medium.

3 Research Approach

The objective of this paper is to examine the genre system found on Facebook during the 2021 Norwegian parliamentary election, compare this with findings from previous elections and to discuss if and how the political parties' Facebook use contribute to realize the objective of reasoned debate, as presented by the constitution.

Data Collection: Data for this study has been collected over four periods: The elections of 2009, 2013, 2017 and 2021. Data for the 2009 study was collected through semi-structured interviews with representatives from the seven political parties that were represented in the parliament before the election (Socialist Left, Labour, Center Party, Liberals, Christian people's party, Conservatives and the Progress Party). In the following elections, follow-up interviews were done electronically, using e-mail or online meetings to confirm findings from 2009. The follow-up interviews had an average of six respondents, bringing the total number of interviews to 25 over the four elections.

Further, social media content (posts, comments and interactions from the pages of the political parties represented in Parliament) during the main campaign period in June to election day in September, has been archived and analysed using Nvivo and Tableau software. For the 2021 election, Facebook's researcher tool CrowdTangle was used for data collection and initial analysis. In addition, statistics from Likealyzer.com, the European Social Survey, and the polling company TNS Gallup has been used to examine trust in media and politics.

Data Analysis: The combination of interviews and content analysis made it possible to compare what informants say with what we can observe happening. This is used to map the genre system in social media political communication. For this study, only Facebook data has been analysed since Facebook remains by far the most used channel in Norwegian politics. The genre systems have been analysed using the 5W1H method presented in the section on genre theory. Of the around 6500 posts collected, a selection has been coded until saturation (no new genres emerging from further study). When no new genres were identified, the remainders of the posts were quickly scanned to see which genre category they matched. Due to space limitations, the findings are presented using the "form/function/content" constructs [24].

Below is an example of how the genres were coded using 5W1H:

Genre: Debate.
Why (purpose): Contribute to a rational deliberative discourse.
When: Continuous, examples found throughout the campaign.
What (content): Text-based. Conversation where participants use rational arguments to discuss a concrete issue.
Who: Politicians, party members, but mostly citizens discussing with each other.
Where: Facebook posts from the various parties.
How (form, tone and style): Encouraging dialogue. Rational arguments. Invites others to reply. Tone is respectful and there are no attacks on person, straw-man type arguments etc.

4 Findings

4.1 Objectives and Channel Use: Same, but Different

In the interviews made in 2009 the political parties agreed on three objectives for political communication in social media': Dialogue with citizens, contributions from citizens, and involvement in party activities. When asked if these objectives remained the same, the parties agreed in 2013. In 2017, they still agreed that these were the overall objectives, but several respondents pointed out that they have evolved and developed a more fine-grained set of strategies, objectives and goals for different channels. In terms of channel use, blogs were popular in 2009, almost gone in 2013. Facebook emerged as the most important channel, and there were some experiments with Instagram. One of the parties said social media communication had been moved from communications to marketing, indicating a stronger shift in objectives than reported. The objectives are presented in Table 1.

Table 1. Political party objectives for social media participation

Objective	Purpose	Form	Content/functionality
Dialogue	Involve citizens in debate about political issues	Encourage dialogue. Open and personal language. Citizen-generated content	Conversation between citizens and politicians
Contribution	Knowledge about citizen concerns	Q&A sessions, Invite voters to share their stories	Encourage contributions and questions from voters
Involvement	Raise funds. Get people to volunteer, Mobilize for action	Competitions, membership forms, information and links to registration sites etc.	Competitions, theme sites, cross-publication
Agenda-setting and informing	Set the agenda for political debate. Inform and educate citizens	Pointed messages designed to get attention and engagement	Informative, short posts. One post – one message. Images and video content

The follow-up interviews in 2021 had similar results to those of 2017. The overall objectives remain the same but have evolved even further and branched out to a multi-channel approach where different social media have different objectives. The overall objectives dialogue, contribution and involvement are supplemented with a new objective called "agenda-setting and informing". Several of the 2021 respondents point out that the perhaps most important objective, at least for Facebook, is to inform (or convince) citizens, and to move the political agenda to issues the party see as their own strong points. While the respondents say the "old" objectives remain equally important, their answers indicate that getting the party's message out is somewhat more important.

Facebook remains the biggest channel in terms of followers and attention, but the audience is getting older, and the parties see other channels becoming more important, especially to reach the younger voters. For example, one respondent said Facebook now was a channel for informing, while they used Instagram as their preferred platform for live dialogue sessions, because Instagram has a more friendly atmosphere. Others mention experimenting with TikTok, as the newest popular social medium. Apart from that,

YouTube and Twitter remain popular, with Twitter more frequently used by individual politicians than the party. But what about how these channels are used? In the next section I present the evolution of communication genres on Facebook, from 2009 to 2021.

4.2 2009–2013: Genre System Evolution

Table 2 presents an overview of the genres identified in 2009 and 2013 and shows how the genre system evolved between elections. A full description can be found in my papers from ePart 2010 and ePart 2014. The third column shows how the genres from 2009 and 2013 evolved into new or similar genres in 2017/2021.

Table 2. Genre system evolution

Genre	2009	2013	2017–2021
Policy comments	Comments from citizens on approved party policy	Present to a lesser degree, movement towards support/-non-support and disgruntlement	Present, but not much
Call for action	Parties call for volunteers, or for action by citizens on specific issues	Still present, and more frequently seen in 2013 after Q&A backlash from 2009	Evolved to contribute
Q&A	Questions from citizens, often unanswered in 2009	Still present, and parties tried to engage more with citizens	Disappeared, users can only comment on posts, not write their own
Appeals to party	Citizens asking what the party intends to do on a specific issue	Evolved into support/non-support and debate	Disappeared, users can only comment on posts, not write their own
Greeting	Greetings and well-wishes on politicians' birthdays etc.	Still present	Still present
Personal accounts	Politicians ask citizens to tell their stories on selected issues	Evolved and split between several genres (call for action, debate, policy comments)	Evolved. Now only between politicians, not asking citizens to share

(*continued*)

Table 2. (*continued*)

Genre	2009	2013	2017–2021
Video responses	Citizens or politicians use video instead of text. Back and forth exchange	Disappeared	Disappeared
Debate	Not present	More users and more activity led to several rational debates on various issues	Some examples, but very little
Support/non-support	Not present	Citizens show support, or lack of support, to parties and party policies	Very much present, both in comments and reactions
Disgruntle	Not present	Sarcastic comments about the party, unpleasant comments about the party and its politicians	Still present, and growing
Link	Not present	Parties link to news articles and other sources. Often accompanied by a short statement or question	Evolved into Slogan genre

4.3 2017–2021: Towards a Stable Genre System and One-Way Communication

In 2017, we saw a marked difference from the election campaign of 2013. While the 2013 election had a lot of interaction, feedback and two-way communication, the 2017 election was a step backwards towards more traditional one-way communication. The same trend is visible in 2021, and there have been no new genres emerging for this latest election.

If you look at Table 2, *policy comment* is the only genre that emerged in 2009 and is still (somewhat) present in 2017 and 2021. In 2013, the new genres *debate, support, non-support, Disgruntlement and Link* appeared. *Link* evolved into the genre *Slogan* in 2017, and the other genres are still present. The most important finding when looking at this evolution is the gradual move from dialogue and feedback to a more simple, one-way form of communication where most of the feedback consists of disgruntled comments, non-support (or to a lesser degree support), and only a few examples of rational debate.

This gradual evolution of genres has given us the following current genre system on Facebook:

We want to is the most commonly used genre from all the parties. The content is directly related to the party program, with statements such as "we want to [do something] because [of some reason]".

We have is only used by the current governing parties. In this genre, the ruling parties present their accomplishments from the last parliamentary session. Sometimes accompanied by the phrase "you know what you have, do you dare vote for something untested". Video and images are frequently used.

Support are common replies to posts on a specific policy, typically short comments stating, "I support/don't support this proposal". However, more and more of these comments are moving towards disgruntlement.

Non-support is frequently used by most parties. In this genre the party attacks the policy and policy consequences of other parties. Political parties have always done this, but the tone is harder than in previous elections. Making fun of the other parties has become a lot more common, as exemplified by the Conservative's image of sun lotion with the text "don't be red this summer, vote Conservative".

Slogan is related to we want to, but in place of concrete policy issues and references to the party program the slogan is more idealistic in nature and is not supported by arguments as to why the statement is true: "We are the best party for young people!" or "Vote for us if you want change".

Personal accounts come in two forms: One is promoting popular politicians in the party, the other is "interviews" with typical voters from large voter groups.

Contribute is where parties ask voters to participate. This can be in the form of Q&A sessions or, more commonly, by asking voters to register for updates, become members of the party or act to support the party.

Society & Context involves parties posting links and updates about current affairs they somehow believe reflects on the values and ideology of the party. For example, the greens post quite a lot about global warming and the conservatives wish people happy pride or post content about the importance of reading.

Experiments is a genre where parties try out different formats of communication, using podcasts or live streaming, giving someone a GoPro to document a day in their lives and similar. Not all parties try this, and the genre is not frequently used. However, this is a sign that there is still some experimentation going on in social media.

Greetings is still a popular way of showing support. Popular politicians celebrating their birthday or other major life event get a lot of attention still.

Disgruntlement is another genre that emerged in 2013 and is sadly growing both in 2017 and 2021. There is a lot of sarcasm and outright hostility towards most of the parties. In fact, most comments and user posts fall into categories arguing for or against the party. This can be interpreted as a sign that polarization is occurring also in Norwegian politics and could indicate that Facebook algorithms show posts by a political party to those who show a preference for opposing parties, in order to generate (negative) engagement.

Debate and **policy comments** are present, but very little compared to the three genres above. There are a few examples of users attempting to start a debate based on evidence, facts and arguments, but most often these posts are taken over by non-supportive or disgruntled comments.

4.4 Growth and Saturation – The Link Between Facebook and Votes

As social media has grown, so has the number of followers and importance of social media as a campaigning channel. In 2017, 23% of the population said they visited Facebook daily to follow the election, on par with the major news media and only beaten by the public broadcaster NRK [13], while in 2009 and 2013 social media was less important as a channel for political news. When using CrowdTangle to examine growth and decline in votes, all the political parties grow by several thousand followers in the month prior to the election. After the election, some of these new followers disappear, while some remain. The number of followers seems to have reached saturation in 2017, with growth between 2017 and 2021 roughly on par with population growth.

There is little correlation between the number of votes a party receive and their number of followers. The Progress party remains the most followed party on Facebook, while they have last almost 270.000 voters from 2009 to 2021. Labour gained 6.000 followers on Facebook, but lost 20.000 votes between 2017 and 2021, while the conservatives gained 20.000 followers, but lost 125.000 votes. The Center party has seen a massive growth in votes, but only limited growth on social media. The Greens have the fourth largest number of followers, but has failed to reach the 4% threshold which gives supplementary seats in parliament in two consecutives elections, mostly because their sympathizers are young, and young people tend to have a lower voter turnout [13]. The only party with a clear correlation between Facebook and voter growth is the far left Red party, which has seen a doubling in both followers and voters between 2017 and 2021. Figure 1 shows the development in followers and votes received between 2009 and 2021 (for Red and Greens the starting year is 2017).

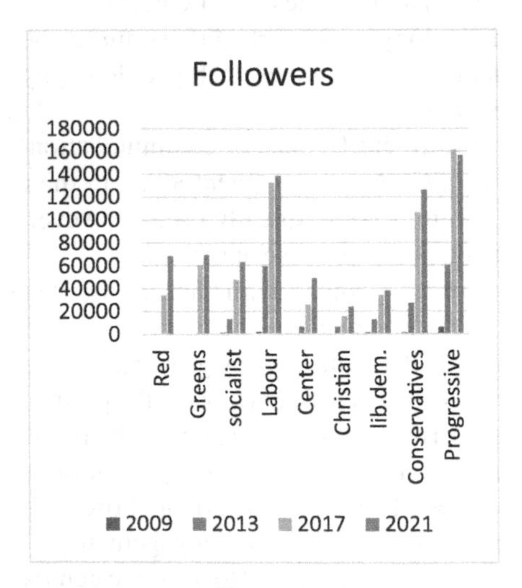

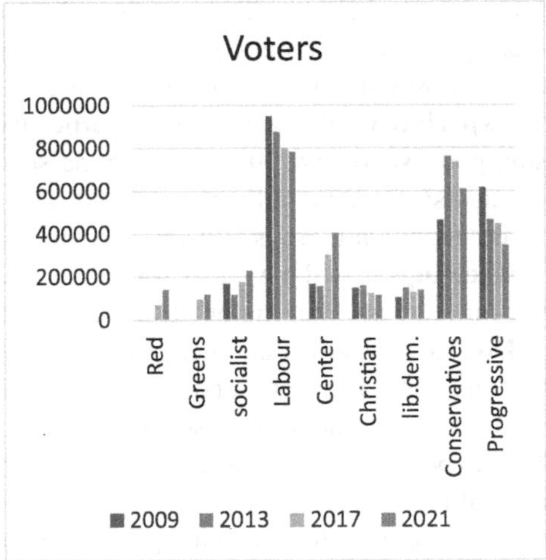

Fig. 1. Followers and votes 2009–2021 (Color figure online)

Demographic variables might be an explanation for this. For example, The Center party has a lot of rural, older voters, who are less likely to be active social media users, while parties such as the liberal democrats and Greens typically attract and urban and younger electorate.

The trend seen in the past two elections also continues: The leaders of the largest parties have more followers than their respective parties, confirming the increased focus on person over party. Politicians' Facebook pages have an interaction rate of 3,9% between July 2021 and the election day in September, while the political parties have an interaction rate of just 1%.

4.5 Interaction and Effect of Genres

While there is little correlation between social media and votes, there is no doubt that social media is engaging voters. In the period from 1 June to election day 18 September, the political parties and party leaders had a total of 3,8 million interactions. An interaction is either a "reaction", what we used to call a like, commenting on, or sharing a post. With 3,9 million people eligible to vote, and 3 million voting, this is an impressive number. However, when drilling further, we see that of these 3,8 million, 362.000 are comments and 125.000 are shares of posts.

The distribution of engagement between party leaders and parties is close to 50/50, but if we also look at other members of parliament, we find several individuals who have a lot of interactions, especially from the two parties who argue about climate and environmental issues, where several politicians from the Greens and the Progress party generate a lot of interactions, as well as nationalist/national romantic posts about Norwegian nature from the agrarian populist Center party.

Looking at the genres for the top posts with the most interactions, there are three genres that emerge as the most effective: the *slogan* genre (statements about what the party wants, with little or no arguments attached), and the *personal account* and *Greeting* genres (personal content, naming other politicians or personal anecdotes). This is of course a reflection of how the Facebook algorithms work, as content receiving attention is spread to more people, receiving even more attention. We also see that most of the top posts have video, image, or links in them, and have few words. Most are limited to two or three short sentences. When scrolling your feed, it is simply easier to stop and like or comment a post with a simple, short message and a picture, compared to a long-complicated text listing a range of arguments. This is also acknowledged by the 2021 interview respondents, who state the importance and difficulty of balancing short posts with a clear stop effect on the one hand, and working with moderation, tact and tone on the other.

Thematically, the posts that generate the most interactions cover either controversial or personal issues. Personal issues range from condolences to recently deceased former politicians to birthday and wedding anniversaries. The reactions here are mostly positive and supporting.

Controversy reflects common dichotomies in Norwegian politics. Climate vs growth, rural vs urban, left vs right ideology and rich vs poor, or the consequences of inequality. Posts cheering climate change mitigation and green transition receive a lot of negative feedback both in comments and angry/laughing reactions, while posts cheering the continued use of fossil fuels, industry over environment and petrol cars receive support. Climate change is perhaps the single issue generating the angriest reactions. Facebook algorithms can be partly to blame, but it might also reflect the fact that the media tends to talk positively about the green transition, while many Norwegians still are not convinced that climate change is an issue, or that Norway, with its renewable energy should do as much as other nations [31]. Other examples include far left vs far-right issues, where one side shouts about socialist danger and the other right-wing extremism, or discussions on public spending (the need for doctors, nurses and teachers) vs public waste of money (heavy bureaucracy and regulations). The growth of the far left Red party, with communist roots, has renewed the left vs right debate in recent years [30], and after eight years of conservative rule the opposition was naturally eager to play on inequality and "the common man vs the rich urban elites". Immigration is traditionally a controversial issue, and one of the top engagement posts discuss immigration, However, this was more of an issue in 2017, after the Syrian refugee crisis.

5 Discussion and Conclusion

All in all, the 2021 election confirms the indications from 2017. Success on Facebook and social media does not guarantee success in the election, even if a lot of media commentators seem to draw a line between the two. Social media is rather part of a complex public sphere ecosystem, where media, social media and people mutually reinforce what we are concerned about. The Red party for example, is surfing on a wave of resentment towards increased inequality, while Labour is suffering from the same reason, as the traditional left-leaning voters move further to the left. On climate change and related issues, we see that even if there is scientific and mostly political consensus about the green transition, many people disagree and voice their concerns on Facebook. Climate change, inequality and rural concerns can of course be seen as part of the current discourse of people feeling disenfranchised and fighting against the elites, but to answer this, more research is needed into the attitudes of active social media users.

There has been a gradual evolution of the genre system found on Facebook. 2009 could be seen as the pilot, with experimentation, few readers but a lot of attention and learning. 2013 showed that the genres were maturing, and there were a lot of attempts at engaging citizens in dialogue, receiving feedback and using the interactive functionalities of social media, while in 2017 and even more so in 2021, we saw a further maturing of genres and no new genres emerged in 2021. The experimental mood from 2013 disappeared, and political parties went back to traditional campaigning using one-way communication, with citizens either cheering or booing from the side line. Some of the explanation for this might be the growth in followers, making it difficult to maintain a deliberative atmosphere in a "space" with people from all walks of life and with little mutual understanding? One interesting avenue for further research could be to examine this further. Is it at all possible to maintain a rational debate following Habermasian ideals in a forum where participants do not share a similar background, do not attempt to understand each other and have no ques about the motivation, reasoning and experiences of the other participants?

All in all, the election campaign of 2021 mostly confirms the findings from 2017. It seems that the genre system on Facebook has matured and stabilized, and while there are several genres in use, those that make use of stop words, controversial slogans and with a focus on personal issues generate the most reactions. In terms of Facebook acting as a public sphere, the answer depends. Applying the criteria of reasoned, rational debate, it is difficult to say that Facebook delivers, and the political parties are more concerned with informing than facilitating debate in this channel. On the other hand, Facebook remains the most popular social medium, where citizens can state their opinion even when it does not strictly adhere to prescribed rules for rational debate. So in conclusion, §100 of the Norwegian constitution is at least partially fulfilled by the political parties' use of Facebook, but there is room for improvement.

References

1. Sæbø, Ø., Rose, J., Nyvang, T.: The role of social networking services in eParticipation. In: Macintosh, A., Tambouris, E. (eds.) ePart 2009. LNCS, vol. 5694, pp. 46–55. Springer, Heidelberg (2009). https://doi.org/10.1007/978-3-642-03781-8_5
2. Medaglia, R.: eParticipation research: moving characterization forward (2006–2011). Gov. Inf. Q. **29**, 346–360 (2012)
3. Habermas, J.: The public sphere. In: Seidman, S. (ed.) Jürgen Habermas on Society and Politics: A reader. Beacon Press, Boston (1989)
4. Arnstein, S.R.: A ladder of citizen participation. J. Am. Inst. Plann. **35**, 216–224 (1969)
5. Ferree, M.M., Gamson, W.A., Gerhards, J., Rucht, D.: Four models of the public sphere in modern democracies. Theory Soc. **31**(3), 289–324 (2002)
6. The Constitution of the Kingdom of Norway – Lovdata. https://lovdata.no/dokument/NLE/lov/1814-05-17
7. Finstad, U.: Falske nyheter i Norge. En kvantitativ undersøkelse av nordmenns oppfatninger om falske nyheter (2021). https://www.duo.uio.no/handle/10852/89833?show=full
8. Kjeldsen, O.B.: Infodemi som smitter En studie om falske nyheter og risikokommunikasjon i Norge under covid-19 pandemien (2021)
9. Frøvoll, G.L.: Ytringsfrihet og polarisering i Norge: En empirisk studie (2021). https://www.duo.uio.no/handle/10852/87330?show=full
10. Sekanina, I.K.: Politisk aktivisme eller ekkokammer? En studie av mannsdebatt på Facebook (2020). https://bora.uib.no/bora-xmlui/handle/1956/23153?show=full
11. Sæbø, Ø., Rose, J., Skiftenes Flak, L.: The shape of eParticipation: characterizing an emerging research area. Gov. Inf. Q. **25**, 400–428 (2008)
12. Mindus, P.: What does E-add to democracy?: Designing an agenda for democracy theory in the information age, pp. 1408–1429(1)AD. https://services.igi-global.com/resolvedoi/resolve.aspx?doi=10.4018/978-1-4666-9461-3.ch073
13. Aardal, B., Bergh, J.: Velgere og valgkamp: En studie av stortingsvalget i 2017. Cappelen Damm Akademisk, Oslo (2019)
14. Østerud, Ø., Selle, P.: Power and democracy in Norway: the transformation of Norwegian politics. Scan. Polit. Stud. **29**, 25–46 (2006)
15. Hestvik, H.: Valgkamp2001.no. Partier, velgere og nye medier. Ny kommunikasjon? In: Aardal, B., Krogstad, A., Narud, H.M. (eds.) I valgkampens hete. Strategisk kommunikasjon og politisk usikkerhet. Universitetsforlaget, Oslo (2001)
16. Fredriksen, A., Alnes, E.: Velgseier med hjelp av Internett (2009)
17. Suh, K.-S., Lee, S., Suh, E.-K., Lee, H., Lee, J.: Online comment moderation policies for deliberative discussion–seed comments and identifiability. J. Assoc. Inf. Syst. **19**, 182–208 (2018)
18. Trenz, H., Eder, K.: The democratizing dynamics of a European public sphere. Eur. J. Soc. Theory **7**, 5–25 (2004)

19. Zhang, W.: Constructing and disseminating subaltern public discourses in China. Javnost Public **13**, 41–64 (2006)
20. Telleria, G.: Revisiting Habermas' public sphere: welcome to the virtual sphere. Acad. Lett. (2021)
21. Miller, C.R.: Genre as social action. Q. J. Speech **70**, 151–167 (1984)
22. Päivärinta, T., Sæbø, Ø.: The Genre System Lens on E-Democracy. Scand. J. Inf. Syst. **20**, 51–82 (2008)
23. Susha, I., Grönlund, Å.: Context clues for the stall of the Citizens' Initiative: lessons for opening up e-participation development practice. Gov. Inf. Q. **31**, 454–465 (2014)
24. Orlikowski, W.J., Yates, J.: Genre repetoire: the structuring of communicative practices in organizations. Adm. Sci. Q. **39**, 541–574 (1994)
25. Yates, J., Orlikowski, W.J.: Genres of organizational communication: a structurational approach to studying communication and media. Acad. Manag. Rev. **17**, 299–326 (1992)
26. Shepherd, M., Watters, C.: The functionality attribute of cybergenres. In: HICSS-32. Proceedings of the 32nd Annual Hawaii International Conference on Systems Sciences 1999, p. 9–pp (1999)
27. Gruber, H.: Genres, media, and recontextualization practices: re-considering basic concepts of genre theory in the age of social media. Internet Pragmat. **2**, 54–82 (2019)
28. Yoshioka, T., Herman, G., Yates, J., Orlikowski, W.: Genre taxonomy. ACM Trans. Inf. Syst. **19**, 431–456 (2001)
29. Yates, J., Orlikowski, W.: Genre systems: structuring interaction through communicative norms. J. Bus. Commun. **39**, 13–35 (2002)
30. Wiborg, Ø.N., Hansen, M.N.: The Scandinavian model during increasing inequality: recent trends in educational attainment, earnings and wealth among Norwegian siblings. Res. Soc. Stratif. Mobil. **56**, 53–63 (2018)
31. Krange, O., Kaltenborn, B.P., Hultman, M.: Cool dudes in Norway: climate change denial among conservative Norwegian men. Environ. Sociol. **5**, 1–11 (2018). https://doi.org/10.1080/23251042.2018.1488516

Legal Informatics

Digitising the Judicial Sector: A Case Study of the Dutch KEI Programme

Lisa Julia Di Natale(✉) and Antonio Cordella

London School of Economics and Political Science, London WC2A 3LY, UK
{l.j.di-natale,a.cordella}@lse.ac.uk

Abstract. Digital justice has long been a buzzword within the judicial community as governments across the world are increasingly transforming their complex judicial bureaucratic structures to build digital institutions. However, the deployment of IT-based systems in the judicial domain has been characterised by high complexities, as most public agents have failed to implement e-justice programmes effectively. The KEI programme of the Netherlands announced in 2012 pertains to such unaccomplished endeavours. Yet, the precise causes and dynamics that led to the withdrawal of this particular digitalisation programme remain to date unknown. This paper investigates the mechanisms and dynamics that governed the integration of the IT architectures during the KEI programme. The paper concludes that culture and structure play a critical role in shaping the negotiation between law and technology, as these have direct effects upon human agency. This paper contributes to existing e-government literature, in particular research on digital justice development, due to the limited exploration that has been seen in the specific field.

1 Introduction

Information Technologies (IT) and novel digital innovations are having transformational effects on information exchange, business conduct and social interaction, and are transfiguring the premises upon which institutional and societal structures have been established. The unprecedented opportunities of technological advancements reach far beyond the commercial sphere, as governments across the globe are increasingly reconstructing complex bureaucratic structures and building digital institutions. The integration of IT into the public sector has transformed the delivery and production of public services, and with that, remodelled the traditional functions of democratic institutions [1, 2]. So far, little attention has been paid to the digitalisation of judicial systems, despite their criticality to nearly all services and activities of the state and society [3–6]. Scholars in the field have explored the emerging changes with the integration of technology into the justice sector and the impact that this has upon established institutional and normative frameworks. Such investigations have revealed that the encounter between law and technology generates a phenomenon that reframes common understandings of digitalisation, calling for richer explorations.

Published by Springer Nature Switzerland AG 2022
R. Krimmer et al. (Eds.): ePart 2022, LNCS 13392, pp. 139–153, 2022.
https://doi.org/10.1007/978-3-031-23213-8_9

Digital justice has long been a buzzword within the judicial community as governments across the world are investing in ITs to innovate and streamline justice systems with the objective of reducing bureaucratisation [7, 8]. However, digitalising the justice sector has been widely characterised as a complex task and most public agents have failed to implement e-justice programmes according to initial development plans. The few empirical studies have highlighted the conflicting logics that must be mediated when assembling law with technology [9]. Whilst documentations on e-justice programmes have been conducted in a range of European countries [6, 10–12], the perhaps most peculiar case remains to date somewhat unexplored: The Dutch Quality and Innovation programme (KEI). In 2012 the Dutch KEI programme was introduced with the vision of digitalising the entire civil and administrative procedures of the judicial sector. The aim of the KEI was to improve efficiency and accessibility of judiciary processes, and to reduce the administrative workload of the courts. The programme was led by the Ministry of Justice (MofJ) as well as Council of Justice (CofJ), and the IT was developed by localised teams At its outset the KEI was a promising programme with carefully thought-out implementation [13] strategies and a seemingly smoothly working development process. However, in 2018, two months before its launch as a comprehensive, operational digital judicial service, the MofJ decided to withdraw the KEI. Whilst it is widely characterised as a 'failed' programme, mainly due to the 'wasted' public funds amounting to 200 million EUR [14], the reasons for this outcome remain unclear. The case of the KEI programme presents an ideal context to dismantle the institutional and normative dynamics that govern the integration of law with technology. The study of the KEI programme offers a unique opportunity to contribute to the limited theoretical understanding that currently exists to this subject.

2 Background

In recent years scholars have shifted their focus away from the purely technological process to more situated studies of technology. Perhaps haunted by the ghost of technological reductionism [15], there has been a strong rise in socio-technical studies of technology capturing the societal, organisational and institutional dimensions which shape the impact of technology on organisation and business processes. Concordantly, a widespread assumption that IT-based systems transform the premises upon which traditional normative and organisational structures are established seems to have diffused amongst scholars [15–17]. Observations have shown that research documenting wider dynamics of IT infrastructures predominantly correspond to the corporate sector [18–20]; whilst institutional effects and implications of technology-based systems remain comparatively less explored within the field of public administration. Yet, various authors can be identified [21–24] to having highlighted the imperative of understanding the societal and institutional landscape into which a specific IT-based system is embedded and have revealed the particular transformative power that technological systems present to the social and organisational framing when implemented in the public domain [3, 9, 25]. The literature that has adopted a more comprehensive perspective seem to attend to the societal and contextual dimensions. Yet, potentially linked to the dominant instrumental perspective, with perhaps some exceptions [3, 9, 26, 27], the dimensions are often treated as causal directions, rather than an interplay. The encounter between technological systems and the complex institutional frameworks, which reach far beyond the situated use of IT [28], should not be limited to strict sequences of interaction nor exhausted at the human-technology interface [21].

2.1 Assemblages

A wide range of concepts have been developed to describe the dynamics that influence the development process of an IT-based systems, especially within strong institutional contexts such as the public domain. The concept of assemblages, originating from Cooper's [29] research on postmodern organisations, is widely used when depicting the techno-institutional encounter. Scholars have extended the notion to studies of IT-development [30], describing imbrications of digital networks and institutional authorities [24]. Lanzara's [27] application of assemblages has added a particular nuance to the subject; exposing the institutional and constitutive valence that such constellations gain. Accordingly, one cannot understand technology as a mere tool for execution as the technical components of assemblages are themselves formative of the institutional environment [26]. The normative and institutional dimensions become wired into the technological circuits, rendering the IT equally critical for execution as well as for enactment and legitimacy [27, 31].

The notion of assemblages clearly departs from the general idea of information infrastructures [20, 32]. Scholars have widely underlined the critical role of the installed base in such discussions, described as the pre-existing and irreversible components upon which IT-based systems are developed [20, 33, 34]. A few scholars have expanded the purely technical understanding of the concept by considering the institutional and organisational component [27]. These perspectives have not only exposed the transformative nature of the phenomenon but have also shed light upon the dynamic interplay from which a novel context-specific assemblage emerges. The process has been described as entailing regulative powers, and institutional constraints and human agency have been identified as the origin from which obstacles arise when developing IT-based systems in the public sector [27, 35]. The relational interdependencies [3] and regulative forces that govern the negotiation appear to still remain under-elaborated within the literature. Scholars seem to have paid more attention to the challenges these present to IT design, where IT development has been described as an "institution building endeavour" [27] demanding both contrasting requirements of robustness and evolvability [36].

Across the research examining ways to overcome the complexities, particularly present within the public sector, a popular theme that emerges is the concept of functional simplification [23]. In order to enable technologies to streamline and standardise institutional objects, administrative procedures and institutional codes must be simplified. Studies have shown that developing an IT system that fully supports highly complex normative structures and processes can technologically be an impossible task [37, 38]. However, empirical studies have also demonstrated that the term functional reduction [27], reducing complexity in line with technical requirements, better captures the process, since simplification, as described by Luhman [23], does not necessarily result in simplified outcomes. The largely agreed upon requirement to reduce pre-established complexities seems to align with the suggestions of human intervention being critical for overcoming regulative tensions and contrasting requirements when developing IT-based systems [32]. Such perspectives explore the transformation of the traditional institutions, which become more technical due to the constraints that IT poses. Contrastingly, the previously reviewed notions instead view the IT artefact as a functional equivalent of the traditional institutions, autonomously producing normative and regulative effects [39], by exploring how the institutional component is inscribed into the technological system [31]. Clearly, two distinct but compatible theoretical reasonings can be identified within the niche literature that dismantles the encounter between IT-based systems and institutions.

2.2 Law and Technology

Independent from which theoretical stream is adopted, when engaging in the development of IT-based systems in the public sector distinct regulative forces and conflicting logics appear to govern the process of digitalisation. On the one hand, such perspectives have exposed the process of technology seizing the active authority of traditional institutions [40]. On the other, notions of the institutional being reduced to the technological portray the institutional dimension itself as becoming a technology of regulation [27]. The digitalisation of the judicial sector has been widely characterised as highly difficult due to its deep formalistic layers [3]. Both formal laws, sustained by procedural codes, and technology, containing its own specifications, are driven by distinct regulating impulses [41] producing a degree of competition between the laws of technology and the technology of law [9]. Embedding IT into a highly regulated context requires novel sets of normative structures and rules [9, 27]. The regulation of technology has been recognised as a specific mode of institutionalisation [40], as the IT must be legitimised and regulated to operate within the legal sector. The first evident consequence observed is that instead of reducing bureaucracy, which is often the aim when digitalising the judicial domain [7, 8], the deployment of technology actually can proliferate laws and regulations [9]. Yet, Kallinikos [39] described technology itself as a "major regulative regime" as it has its own normative authority, with the power to regulate and constraint institutional procedures. Whilst both the law and the technology engage in normativity, the modes of regulation and intrinsic operating logics are strikingly different: technology orients itself by the outcome and is thus judged theologically [42]; law on the other hand defines legitimacy and is judged deontologically [9] – meaning that digital justice systems must be judged both for their legitimacy and effect [43]. The discussions overall indicate that the techno-legal encounter can be understood as a competitive negotiation between regulative regimes, in which technology and law strive to civilise each other [31].

Mediating Regulative Regimes. The extent to which IT-based systems affect the accessibility, legitimacy, and legality of judicial systems has been recognised as a great challenge when introducing technology into the judicial sector [38]. An even greater conflict in logics and complexities appears to be observed than the implementation of IT-based systems into more agile normative environments. Accordingly, Contini and Mohr [9] have conducted a range of empirical studies assessing the development process of digital justice systems, in which the mediation of the regimes [44] is widely marked as crucial to the integration of law and technology. Across such research, a wider debate emerges regarding the regulation of either the law or the technology for overcoming the resulting complexities. The most pronounced discourse seems to surround the ex-ante regulation of technology, where scholar described that when law has been used to specify the technological conditions ex-ante, the technology proliferated to the extent that it was unusable and the law unintelligible [9]. Following this diagnosis, suggestions have been made that the correct order of regulation must be identified to conciliate the negotiation between law and technology. The opposite solution would be post-facto legitimisation, which some have recognised as critical since solely the enactment and usage of the system can demonstrate if the implemented IT-solution can meet the requirements [45, 46]. Hence, ex-post regulation would provide for an adequate legal framework tailored to the IT-system, reducing the risk of unnecessarily proliferating either component, by adhering to the perhaps more compelling pre-programmed requirements of technology. As expressed by Lessing [31]: code is law. However, regulating the law through post-facto legitimisation equally can create complexities as the developed IT - system might not meet the necessary legal requirements for it to operate [9]. Whilst such debates have been central to the realm of e-justice, the dissensus regarding the mediation of the techno-legal regulative regimes leaves unanswered questions within the literature.

Missing Component: Human Agency. Overall, the digitalisation of justice systems appears to be a field that calls for further exploration. Whilst the niche literature exposes the regulative regimes that both law and technology constitute and indirectly indicates that human intervention results as critical for the mediation of the conflicting logics, the human agency component and the effect of their actions upon the techno-legal negotiation seem to not be thoroughly dismantled. Yet, research has demonstrated that integrating technology with institutions interlinks a wide range of actors, from which new constellations of human agency emerge. New actors and workflows have been highlighted to constitute a potential new source of instability: human agency introduces its own regulative regime, carrying cultures and constraints, which equally can proliferate with its expansion [9]. Studies on the transition into Enterprise IT have provided for some valuable insight into human agency, highlighting that measures such as decision-making rights must be established ex-ante for achieving successful transformations [47]. Hence, managerial practices determine the spheres of agency and responsibility appropriate to the specific functional structure and assemblage [9], which overall endows human agency with perhaps even greater, or at least just as determining regulative powers to the regimes of law and technology.

3 Methodology

Considering the comparatively unexplored dimension of human agency and its concomitant critical dynamics the research question that we posed is how the culture and structure of the organisation involved in developing IT-systems in the judicial domain shape the negotiation between law and technology?

To answer the research question, this paper adopts a qualitative research design in the form of a case-study analysis, as it enables the identification of underlying mechanisms and allows for an in-depth understanding of complex dynamics. Furthermore, the territoriality of the judicial sector and the context-specificities of the phenomenon can be accounted for and examined more thoroughly with the study of a particular case. Building on explanatory case studies [48, 49] the aim of this paper is to identify themes and mechanisms that can be transposed onto a general sphere, taking into consideration that integrating IT into the judicial domain and overall public sector is an agenda that most governments across the world have been introducing.

The deep formalistic layers that have characterised the judicial sector as a highly difficult domain to digitalise presented itself as a resourceful and challenging field of investigation. The KEI case was selected due to its unexplored nature that has left a range of questions unanswered regarding the causes behind its unexpected outcome. This case also exhibits the conflicts that emerge between law and technology, as well as dynamics pertaining to human agency.

3.1 Data Collection Strategy

The data collection strategy is based on methodological triangulation [50, 51] with the intent to reduce intrinsic bias that single-method studies contain and to better dismantle the phenomenon under study. Semi-structured, in-depth interviews were conducted of the two most prominent figures of the KEI programme, namely, the Product Owner (PO) and the Programme Director (PD). Interpretations and information are elicited through a combined technique of guiding conversation and asking targeted questions [52]. The two interviews conducted with the PO will be annotated as I.1.R and I.2.R; the interview with the PD as I.1.S. Several documents, such as a report released by the Judicial Council in 2014 [53], advised as insightful sources by members of the KEI, are used to complement the findings. Combining the interviews with the documents serves to expose the specific dynamics that shaped the unfolding of the KEI programme in an answer to the research question.

4 Case Study

Already in the nineties the Judiciary of the Netherlands feared falling behind other public functions and initiated a renewal program, ranging from the introduction of webservices to the development of an overarching quality system. The Dutch judicial structure constituted a dual system in which the president of court oversaw judicial matters, and the minister of Justice was responsible for court management. In 2002 this structure was replaced by integrally responsible boards to govern the courts and a Council for the Judiciary to manage the overall Judiciary [53], including budget distribution and advice on new legislations. Judicial Independence was maintained: judges are independent from the president of the courts, giving them authority over cases, procedures, and their overall modes of operation. The Council continued the renewal initiatives and implemented several quality standards and release of output-based funds to support the gradual expansion of digital access to the courts. The Netherlands is divided into eleven district courts, four courts of appeal and one Supreme Court and each court accounts for its own working process. In 2012 several courts were merged to unify the diverse administrative procedures that exist across Dutch courts, which created a wide gap between procedural rules and practice. However, a single operating system was not established. The harmonisation of the administrative processes and the fusion of courts was met with a strong resistance as judges wanted to maintain their own administrative processes. As each court is based in two to three distinct locations, this meant that the judicial system accounted for 50 distinct operating processes. The 2014 advisory of the Council of State highlights that suggestions to omit the distinction between claims and requests procedures in civil justice were not followed due to resistance from legal practice, which increased the complications and costs of digitalisation [54].

4.1 Quality and Innovation Programme (KEI)

The Netherlands sought to digitise the entire civil and administrative procedures through the Quality and Innovation programme (KEI) announced in 2012. This programme aimed to make the judiciary processes quicker and more accessible, and to reduce the administrative workload of the courts. The Council aimed to realise new processes by simplifying and standardizing procedures [53]. The main feature of the programme was the development of a novel digital procedure for civil and administrative law, as the Dutch judiciary sought to shorten procedures at the court. The key words with which the KEI was introduced were: speed, quality, simplicity, uniformity and transparency [53]. Ultimately, the programme had two objectives: digitalising the judiciary and economising on the judiciary budget. Part of the specific aims of the KEI was to reduce the administrative workload by 43% within three years with the digitalisation of all case law processes [55]. The Ministry of Justice set to finalise the entire digitalisation programme within four years, in accordance with the cabinet period.

The programme consisted of three components: supervision, administration, civil, of which each had one or more projects. The KEI legislation was part of the programme, which mainly consisted of amendments to the Code of Civil Procedures and to General Administration Law Act to enable and enforce digital litigation, amongst other. The legislation was adopted unanimously in July 2016, in parallel with the IT development. For instance, as Article 113 of the Code of Civil Procedures, allowing for the involvement of the opposing party prior to court proceedings, was added in a later phase of the legislative process, the design for digital access in civil matters had to be drastically revised [55]. Several changes to the Dutch legal framework were undertaken throughout the KEI programme to legitimise the streamlined digital information exchange. This programme was led by the Ministry of Justice of the Netherlands (MofJ) and the Council of Justice (CofJ), and the IT was developed by localised teams using an agile product development method [13]. The MofJ oversaw the new procedural legislation and provided the funding, and the CofJ had the decision-making power over the IT implementation. The KEI had a programme board consisting of two Council members, an IT Director and a Programme Director. Over the course of the programme, the governance structure increased in complexity, with the greatest amplification documented in 2014, as programme managers were hired, and individual project boards were implemented. A group of judges was assembled who held the authority over the working process of the KEI, as well as the area and degree of digitalisation.

At its outset, the KEI was a promising programme with carefully thought-out implementation strategies and a seemingly smoothly working development process. However, in April 2018, two months before its launch as a comprehensive operational judicial service, the MofJ decided to abruptly withdraw the programme. Accordingly, the expenses of the programme stood far higher than anticipated and the implementation and development period equally surpassed the originally established timeframe. The programme has since been widely characterised as a 'failure', mainly due to its investment amounting to 200 million EUR [14]. On the other hand, the parliamentary hearings on government IT-projects have clarified that an IT-project can only be considered unsuccessful when the developed software is not implemented – 20.000 asylum cases, 80 percent of bankruptcies, and 4000 commercial claims have been handled with systems developed under the KEI (I.1.R) – suggesting that the withdrawal of the KEI did not occur uncontested:

The KEI became the victim of double aims: quality and economising (I.1.S.). Not only is the idea that IT will lead to cost reduction a common fallacy (I.1.R.), but the case of the KEI demonstrated that it also promoted resistance towards the introduction of technology, hence, having a direct effect upon relevant actors. Reducing the administrative workload by 43%, effectively meant that the administration would have to cut half of its staff, which was received with strong opposition (I.2.R.). The support of the judges turned out crucial to the integration of IT. Focusing the aims on quality and access could have promoted both internal and external support; emphasising the benefits of digitalisation has long-lasting effects (I.1.S.).

5 Case Analysis

5.1 Regulative Regimes

Increased complexity inevitably proliferates risk [30]; the short answer to the question on how to effectively develop and implement court IT is to '*simplify, simplify, simplify*' (I.2.R.). The digitalisation process of the Dutch judiciary resulted in a proliferation of law to accommodate the needs of the IT (I.1.R.). Procedural changes and modifications of the digital processes rendered the adoption for legal practitioners much more difficult, as they had to become acquainted with both the digital system and the new procedures. The cumbersome legal and procedural reformations directly affected human agency and complicated the digitalisation. '*Every time something came from the Top and I thought "this is not simplification" I would turn it around until it was a workable thing to do (...) there was no consistent body of requirements, requirements themselves changed (...) what made life complicated was this tandem with the legislation*' (I.1.R).

The IT system was designed to simplify procedures, however any changes in the procedural requirements meant that the system had to be revised or redesigned from scratch. The digital access of civil cases was one of such major reconfigurations (I.1.R.). Inscribing judicial procedures into IT-based systems requires novel procedural and legal requirements, which in turn, the technological installation must sustain to be legitimised; resulting in an increase of both technical and legal complexities. This not only exacerbated the cost and time of the KEI programme, but also intensified the evident tensions between law and technology.

'*IT cannot tell us what to do*', and '*lawyers cannot tell IT how it should work, as that would result as very costly*' (I.1.R.).

The conflicting regulatory dynamics that governed the KEI programme align with the notion that law and technology engage in a competitive negotiation, where one strives to *civilise* the other [31].

The digitisation of procedural practices also generates a further layer of complication to the judicial practices [56].

'*Every court thinks that their method of dealing with cases is the best administrative procedure (...) because this is an area where the law is not very leading, there is a lot of discretion in how to deal with it; this is where the problem starts*' (I.1.S.).

The digital optimism IT standardisation enforces was not completely misguided, however, '*if you do not have any harmonised process at all, it is overstretched to think that you can harmonise them by digitalisation*' (I.1.S.). This multitude of discretionary procedures would have required the design of separate executive information systems linked to distinct processes and IT systems – an excessively complex task for designers. The constraining power of technology not only requires mediation with formal legal requirements [39] also working processes must be harmonised prior to digitalisation (I.1.S.). Moreover, the reluctance of judges to abandon habitual procedures combined with their authoritative power impeded the digitalisation process of the KEI (I.2.R.; I.1.S.). The executive actors ultimately determined the negotiation between law and technology.

5.2 Structure

As seen thus far, human agency played a critical role in the trajectory of the KEI, opposing both legal and technological demands and autonomously producing normative and regulative effects by imposing procedural conduct. The objection to digitalisation – originating from the desire to conserve the traditional environment – posed by the judges overruled the objectives of the KEI leaders. This *bargaining position* (I.1.S.) is reflective of the traditional structure of the Dutch Judiciary, which was '*copied onto this digitalisation project and resulted as disastrous*' (I.1.S.). This structure functionally enabled judges to oppose the unified operating process, which was essential to the development of the IT, by giving them the formal authority to execute their discretionary power and maintain their individual customary working processes.

'*It's like building a house: you have influence over the builder prior to a contract, but once you have signed a contract and new events occur, you have almost no influence (...) and we all know that the extra works for uncertain events are usually far more expensive*' (I.1.S.).

Suggestions have been made that the top of the governance-pyramid must contain a single actor responsible for content, money and time, with the power to override the decisions of the judges and objectively evaluate what must and can be digitalised. The following level would see the PO receive instructions and requirements from a group of judges, with the power to refer to the Top in case of compromises (I.1.S.).

'*There should be no undue influence on the content of decisions of the Judiciary, but in digitalisation this is not only a nightmare; it is a receipt for failure*' (I.1.S.).

Such an order would distance the digitalisation process from traditional normative settings; if change is desired structure and durability of old practices must be abolished and new arrangements designed [27]. Yet, instead of suspending old practices, the KEI programme underwent a wide proliferation in structure and actors, which created new sources of instability [57]. In the initial stages, the structure was remarkably simple: the PO had a direct line with the PD, which facilitated the distribution of information and communication. In the final stages, there were three hierarchical levels between the PO and the PD (I.1., 2.R; I.1.S). The new constellations of human agents increased managerial and procedural complexities [9].

'*A project manager whose domain was unsure reported on what my project had been doing (...) it would have been so much easier if I had done the reporting myself*' (I.1.R.).

5.3 Culture

The KEI case exhibits the constraining effect that structure can have upon digitalisation. Moreover, it exposed how actors can be caught between the shadows of the past and the vision of the future [58]; a seemingly silent dynamic critical to the digitalisation process of the Dutch justice system.

The strong counteracting pressures combined with judges' effective authority, which orchestrated the mediation between the judicial and the digital regime, brings to the foreground the role that institutional culture play in defining the trajectory of digitization project in the judiciary [3]. The conservative nature of judges defined the direction of the techno-legal negotiation: conserving the traditional environment and dismissing the path towards digital innovation.

Actors have the power to shape the ambivalent nature of institutional culture, using it as trap or resource to innovation, depending intrinsic characteristics of the existing culture [27]. Our case revealed that court-culture and the long-established fundaments of the Judiciary are the origins from which much of the complications emerged.

'*The IT team had decided to utilise agile development methodology, the opposite to bureaucratic (...) It's not like building a standard house that you know where all door knobs will be (...) this was something so new that experimenting was the only way forward and the agile methodology is very good for that (...) the Top did not realise that that's what needed to be done*' (I.1.R.).

The clash between the bureaucratic court culture and the agile principles of the IT team increased tensions between the two bodies. For instance, methods of reporting were strikingly different: the project team used Demos whilst the judicial body demanded extensive documentations (I.2.R). Agility allows for quick responses, which is essential when developing an IT-based system constrained by requirements, costs, and time [55]. Bureaucratic working cultures are not reflective of agile principles: '*I don't exaggerate, if they (judges) act swiftly, they take a decision in four weeks*' (I.1.S.). Every two weeks a team working on the digitalisation costed one million EUR, interrupting the project for four would cost two million EUR, which judges countered with "*this is how we work*" – '*and they are right, but it is not how you can develop IT*' (I.1.S.). These conflicting cultures governed the working dynamics of the KEI and consequently raised the difficulties of reaching the digitalisation aims. '*Legal thinking is about guilt and responsibility, whereas agile methodology focuses on problem and solution*' (I.2.R.). The inherent contrast between law and technology also manifests as a cultural conflict, as shown in our case.

'These two cultures had difficulty working together (…) the agile way was extremely effective but made them very insecure as it was not like the routine they knew' (1.2.R.).

Maintaining the traditional working culture not only increased complications, but also further exposed the conservative nature of the judicial body. Yet, contrary to previous suggestions, authority over a digitalisation process should always remain with the domain that undergoes digitalisation [55].

'Tech people should not decide on how the judiciary works, but the judiciary should learn to understand how they could work with technology' (I.2.R.).

This is where the intrinsic features of the existing institutional and normative framework come into play: the role of the Judiciary in society cannot be compromised with digitalisation. The monopoly of the Judiciary over laws and regulations requires citizens' trust, which is why the institution holds Judicial Independence [54]. In order to maintain independence, the Judiciary must have the power to decide over work processes, and if these are digitalised then they must hold the authority over the integration of IT (I.1.R.). Considering the findings thus far, the highly normative and regulative environment of the judicial sector along with its fundamental judicial principles renders it an inherently complex realm to digitalise, since the long-established institutional structure itself produces conflicts with technology. The authority should thus remain with the Judiciary, but the agile culture necessary for developing an IT-based system must be embraced. The cultural constraints that judges pose upon modernisation have been reported the review committee of the Judiciary [59]: The '*inward-looking*' culture has an obstructive effect upon society, which changes at a much faster pace than the Judiciary. The '*change fatigue*' produces obstacles to evolution and the transition into a modern institution is a sine qua condition for that the Judiciary retains its authoritative power. As change produces instability and transmits uncertainty, old practices and structures quickly become more salient, which, in line with systems dynamic theory, may render change and innovation efforts self-defeating [31, 60]. The traditionalist court culture and the premises upon which the Judiciary is established are the key sources from which the conflicts with the digitalisation process during the KEI emerged in our analysis, aligning with Contini and Mohr's [9] notion that actors and their cultures constitute an own regulative regime. This conservative court culture produced obstructing effects upon digitalisation, shaping the negotiation between law and technology – which, due to the authoritative position of the judges, saw conservation win.

Our case shows that culture and structure result as the determining components of integrating IT into the judicial sector.

6 Findings

Overall, this research has contributed to existing theoretical notions by emphasising the role of human agency in the digitalisation process [26, 27, 39]. The literature has revealed the cross-cutting environment of regulative regimes into which an IT - based system is embedded and how law, technology and human agency are positioned accordingly [25, 61]. Revisiting Kallinikos [39], the regulatory forces of law and technology are in fact not only directed at one another; they shape human agency by engraving particular sequences of action and also shape cognitive processes of perceptions, preferences and organisational norms. This research has uncovered how human agency has been proven as the most critical dimension to the digitalisation of the judiciary. The concomitant components of human agency that have shaped the negotiation between law and technology and thus determined the digitalisation process within the specific case-study are culture and structure. Culture has been identified as the variable that defines the negotiation and with that the outcome of the techno-legal encounter; and structure as the component that functionally enables the orchestration of the negotiation. Whilst both components are not necessarily dependent upon each other, in the examined case structure was defined by culture and the regulative effects of culture was functionally enabled by structure – opposite effects and independence may equally exist. In the specific case-study judges dictated the negotiation between law and technology, which was defined by their culture and enabled by the traditional structure of the Dutch Judiciary. Chiefly the culture of discretionary power of legal practitioners who placed counteracting pressures on the digitalisation process, due to the fundamental customary change that this would entail, strongly determined the techno-legal negotiation. Examining the structure of the KEI revealed that the traditional order of the Dutch Judiciary was transposed onto the programme, which with the principle of Judicial Independence enabled judges to dictate the integration process of IT and effectuate their objection. Furthermore, strong cultural conflicts have been documented, both regarding the bureaucratic and agile operating methods and the conservative culture that dominates the judicial sector, which greatly expanded the existing limited knowledge on human agency and its concomitant components in relation to techno-legal innovation. Hence, the answer that this investigation proposes to the stipulated research question is: the actions of the agents involved in developing an IT-system are defined by culture and the degree of effect is determined by structure, which is how these shape the negotiation between law and technology.

7 Conclusion

The purpose of this paper was to explore the dynamics and regimes that govern the digitalisation process of the judicial domain. This research endeavoured to examine the way in which actors involved in integrating an IT-based system into the justice sector shape the negotiation between the law and technology.

Investigating the Dutch KEI programme has exposed the power struggle between law and technology and culture and structure. The case has revealed the strong constraints that institutional culture poses on the negotiation between law and technology. These empirical findings have led to the conclusion that culture determines the ease of the transition into the digital realm, hence defining the techno-legal negotiation, and structure functionally enables the orchestration of this process. This research has been conducted in a specific case and context. Further research could explore the role of human agency and the techno-institutional interplay across different cases and less formalistic sectors. Overall, this study strengthens the idea that the cross-cutting environment of distinct regulative regimes into which an IT-system is embedded must be recognised and the criticality of human agency not underestimated. Taken together, this paper has demonstrated that great efforts are needed on behalf of the Dutch Judiciary to ensure an efficient transition into the digital domain in order to meet the exigencies of our fast-paced society. Digitalisation is a phenomenon of great underestimations and misconceptions, embarking on this quest should be seen as a gradual and long-term endeavour.

References

1. Castells, M., Cardoso, G.: The Network Society: From Knowledge to Policy. Center for Transatlantic Relations, Paul H. Nitze School of Advanced (2005)
2. Fountain, J.: Central issues in the political development of the virtual state. In: The Network Society from Knowledge to Policy, vol. 149 (2005)
3. Contini, F., Cordella, A.: Assembling law and technology in the public sector: the case of e-justice reforms. In: Proceedings of the 16th Annual International Conference on Digital Government Research, pp. 124–132 (2015)
4. Fragale Filho, R.: The use of ICT in Brazilian courts. Electron. J. e-Gov. **7**(4), 339–348 (2009)
5. Contini, F., Lanzara, G.F.: The Circulation of Agency in E-Justice: Interoperability and Infrastructures for European. Springer, Dordrecht (2014). https://doi.org/10.1007/978-94-007-7525-1
6. Contini, F., Cordella, A.: Information system and information infastructure deployment: the challenge of the italian ejustice approach (2004)
7. Cordella, A., Willcocks, L.P.: ICT, marketisation and bureaucracy in the UK public sector: critique and reappraisal. In: Contini, F., Lanzara, G.F. (eds.) ICT and Innovation in the Public Sector, pp. 88–111. Springer, Heidelberg (2009). https://doi.org/10.1057/9780230227293_5
8. Velicogna, M., Contini, F.: Assemblage-in-the-making: developing the e-services for the justice of the peace office in Italy. In: Contini, F., Lanzara, G.F. (eds.) ICT and Innovation in the Public Sector, pp. 211–243. Springer, London (2009). https://doi.org/10.1057/978023 0227293_9
9. Mohr, R., Contini, F.: Reassembling the legal: 'the wonders of modern science' in court-related proceedings. Griffith Law Rev. **20**(4), 994–1019 (2011)
10. Koch, S., Bernroider, E.: Aligning ICT and legal frameworks in Austria's e-bureaucracy: from mainframe to the Internet. In: Contini, F., Lanzara, G.F. (eds.) ICT and Innovation in the Public Sector, pp. 147–173. Springer, London (2009). https://doi.org/10.1057/978023022 7293_7
11. Fabri, M.: E-justice in Finland and in Italy: enabling versus constraining models. In: Contini, F., Lanzara, G.F. (eds.) ICT and Innovation in the Public Sector, pp. 115–146. Springer, London (2009). https://doi.org/10.1057/9780230227293_6
12. Carnevali, D., Resca, A.: The civil trial online (TOL): a true experience of E-justice in Italy. In: 2012 AA. VV. "Building Interoperability for European Civil Proceedings online (2012)
13. Reiling, D.: Technology for justice: how information technology can support judicial reform (2016)
14. Schmidt, A.H., Zhang, K.: Agent-based modelling: a new tool for legal requirements engineering: introduction and use case (KEI). Eur. Q. Polit. Attitudes Mentalities **8**(1), 1–21 (2019)
15. Kallinikos, J.: The order of technology: complexity and control in a connected world. Inf. Organ. **15**(3), 185–202 (2005)
16. Castells, M.: The Rise of the Network Society. Wiley-Blackwell, Oxford (2010)
17. DiMaggio, P.: The Twenty-First-Century Firm: Changing Economic Organization in International Perspective. Princeton University Press (2003)
18. Weill, P., Broadbent, M.: Leveraging the New Infrastructure: How Market Leaders Capitalize on Information Technology. Harvard Business Press (1998)

19. Monteiro, E., Hanseth, O.: Developing corporate infrastructure: implications for international standardization. In: Paper to 1st IEEE Conference on Standardisation and Innovation in Information Technology, Aachen, September 1999
20. Ciborra, C., et al.: From Control to Drift: The Dynamics of Corporate Information Infrastructures. Oxford University Press on Demand (2000)
21. Borgmann, A.: Crossing the Postmodern Divide. University of Chicago Press (2013)
22. Hughes, T.P.: Networks of Power: Electrification in Western Society, 1880–1930. JHU Press (1993)
23. Luhmann, N.: Membership and motives in social systems. Syst. Res. **13**(3), 341–348 (1996)
24. Sassen, S.: Territory, Authority, Rights. Princeton University Press (2008)
25. Gualdi, F.: Artificial intelligence and decision-making: the question of accountability. In: Cordella, A. (ed.) Proceedings of the 54th Hawaii International Conference on System Sciences, p. 2297 (2021)
26. Ciborra, C.U., Lanzara, G.F.: Formative contexts and information technology: understanding the dynamics of innovation in organizations. Account. Manag. Inf. Technol. **4**(2), 61–86 (1994)
27. Lanzara, G.F.: Building digital institutions: ICT and the rise of assemblages in government. In: Contini, F., Lanzara, G.F. (eds.) ICT and Innovation in the Public Sector, pp. 9–48. Springer, London (2009). https://doi.org/10.1057/9780230227293_2
28. Misa, T.J., Brey, P., Feenberg, A.: Modernity and Technology. MIT Press, Cambridge (2003)
29. Cooper, R.: Assemblage notes (1998)
30. Ciborra, C.U.: Interpreting e-government and development: efficiency, transparency or governance at a distance? In: Avgerou, C., Lanzara, G.F., Willcocks, L.P. (eds.) Bricolage, Care and Information, pp. 90–110. Springer, London (2009). https://doi.org/10.1057/9780230250611_5
31. Lessig, L.: Code: And Other Laws of Cyberspace. ReadHowYouWant.com (2009)
32. Hanseth, O.: Information technology as infrastructure (1996)
33. Chae, B., Lanzara, G.F.: Self-destructive dynamics in large-scale technochange and some ways of counteracting it. Inf. Technol. People **19**(1), 74–97 (2006)
34. Hanseth, O., Lundberg, N.: Designing work oriented infrastructures. Comput. Support. Coop. Work (CSCW) **10**(3), 347–372 (2001). https://doi.org/10.1023/A:1012727708439
35. Hanseth, O.: The economics of standards. In: From Control to Drift: The Dynamics of Corporate Information Infrastructures, pp. 56–70 (2000)
36. Wagner, A.: Robustness, evolvability, and neutrality. FEBS Lett. **579**(8), 1772–1778 (2005)
37. Contini, F., Lanzara, G.: ICT and Innovation in the Public Sector: European Studies in the Making of e-Government. Springer, London (2008). https://doi.org/10.1057/9780230227293
38. Cordella, A., Contini, F. (ed.): Digital Technologies for Better Justice: A Toolkit for Action. Inter-American Development Bank (2020)
39. Kallinikos, J.: The regulative regime of technology. In: Contini, F., Lanzara, G.F. (eds.) ICT and Innovation in the Public Sector, pp. 66–87. Springer, London (2009). https://doi.org/10.1057/9780230227293_4
40. Czarniawska, B.: How institutions are inscribed in technical objects and what it may mean in the case of the Internet. In: Contini, F., Lanzara, G.F. (eds.) ICT and Innovation in the Public Sector, pp. 49–65. Springer, London (2009). https://doi.org/10.1057/9780230227293_3
41. Hildebrandt, M.: A vision of ambient law. In: Regulating Technologies, pp. 175–191 (2008)
42. Weick, K.E.: Technology as equivoque: sensemaking in new technologies. In: Technology and Organizations (1990)
43. Kelsen, H.: General theory of norms (1990)

44. Hanseth, O., Lyytinen, K.: Design theory for dynamic complexity in information infrastructures: the case of building internet. In: Willcocks, L.P., Sauer, C., Lacity, M.C. (eds.) Enacting Research Methods in Information Systems, pp. 104–142. Springer, Cham (2016). https://doi.org/10.1007/978-3-319-29272-4_4
45. Wallace, A.: Virtual justice in the bush: the use of court technology in remote and regional Australia. JL Inf. Sci. **19**, 1 (2008)
46. Henning, F., Ng, G.Y.: The challenge of collaboration–ICT implementation networks in courts in The Netherlands. Transylvanian Rev. Adm. Sci. **5**(28), 27–44 (2009)
47. McAffee, T.B.: Restoring the lost world of classical legal thought: the presumption in favor of liberty over law and the court over the constitution. U. Cin. L. Rev. **75**, 1499 (2006)
48. Benbasat, I., Goldstein, D.K., Mead, M.: The case research strategy in studies of information systems. MIS Q. **11**(3), 369–386 (1987)
49. Yin, R.K.: Design and methods. Case Study Research **3**(9.2) (2003)
50. Triangulation, D.S.: The use of triangulation in qualitative research. In: Oncol Nurs Forum 2014, vol. 5, pp. 545–547 (2014)
51. Campbell, D.T., Fiske, D.W.: Convergent and discriminant validation by the multitrait-multimethod matrix. Psychol. Bull. **56**(2), 81 (1959)
52. DeCarlo, M.: 13.2 Qualitative interview techniques. Scientific Inquiry in Social Work (2018)
53. de Groot-van Leeuwen Leny, E.: Judicial reform in The Netherlands. In: Rechtspraak, D. (ed.) Scientific Magazine for the Judiciary Organisation of The Netherlands. Sdu Utigvers (2014)
54. (Dory) Reiling, A.D., (Guus) Harten, G.W.J., (Erik) Boerma, F.H.E., (Michiel) Boer, M.Ch.: De Legal Tech van de rechtspraak. https://home.hccnet.nl/a.d.reiling/html/De%20Legal%20Tech%20van%20de%20rechtspraak%20v%202.pdf?msclkid=439d0d89a6e711ecae13ca2471a20c89. Accessed 08 Mar 2020
55. Reiling, D.: Court information technology: hypes, hopes and dreams. In: Kramer, X., Biard, A., Hoevenaars, J., Themeli, E. (eds.) New Pathways to Civil Justice in Europe, pp. 43–60. Springer, Cham (2021). https://doi.org/10.1007/978-3-030-66637-8_3
56. Bovens, M., Zouridis, S.: From street-level to system-level bureaucracies: how information and communication technology is transforming administrative discretion and constitutional control. Public Adm. Rev. **62**(2), 174–184 (2002)
57. Contini, F.: ICT, assemblages and institutional contexts: understanding multiple development paths. In: Contini, F., Lanzara, G.F. (eds.) ICT and Innovation in the Public Sector, pp. 244–271. Springer, London (2009). https://doi.org/10.1057/9780230227293_10
58. Kuran, T.: The tenacious past: theories of personal and collective conservatism. J. Econ. Behav. Organ. **10**(2), 143–171 (1988)
59. Gerechten, R.V.: Goede rechtspraak, sterke rechtsstaat. In: Rechtspraak, D. (ed.) (2018)
60. Forrester, J.W.: Counterintuitive behavior of social systems. Theory Decis. **2**(2), 109–140 (1971). https://doi.org/10.1007/BF00148991
61. Kallinikos, J.: The regulative regime of technology. In: Governing Through Technology, pp. 12–33. Palgrave Macmillan, London (2011)

A Song of Digitization and Law

Design Requirements for a Digitization Check of the Legislative Process

Michael Koddebusch[1(✉)], Sebastian Halsbenning[1], Lennart Laude[2], Vivien Voss[2], and Jörg Becker[1]

[1] ERCIS, University of Münster, Münster, Germany
{koddebusch,halsbenning,becker}@ercis.uni-muenster.de
[2] Lorenz-von-Stein Institut, Kiel University, Kiel, Germany
{llaude,vvoss}@lvstein.uni-kiel.de

Abstract. The digital transformation of the public sector needs laws that are designed to foster digitization. Scholars have long investigated different aspects of the legislative process and how to enhance it with regard to digital meaningfulness. However, to date, there is scarce research to expand these efforts beyond the legislator and consider public authorities charged with the enforcement of laws. The study employs a qualitative research approach to bridge this gap. We conducted nine expert interviews, aiming to introduce design requirements for a digitization check that considers public authorities enacting and enforcing legislation. Fourteen design requirements were established, which are essential for implementing a digitization check. Complementary to the requirements, subjects to be covered by the check are presented. The results of this study enable practitioners to the legislative process and call for scholars to deepen their understanding of digital-ready legislation.

Keywords: Digital-ready legislation · Design requirements · Digitization check · e-Government · Public sector digitization

1 Motivation and Introduction

The ever-faster pace of the digital transformation, meaning the usage of digital technologies creating disruptions, nowadays concerns all parts of our societal coexistence [27]. Whereas significant effects of this disruptive process have already become increasingly evident in the economy and society, the public sector does not yet show the same level of transformation [12]. Scholars and practitioners have aligned on the understanding that the digital transformation poses significant challenges to the public sector [3,7], but, at the same time, offers an approach to solving other present issues [16], e.g., the increasing workload for a continuously decreasing workforce. The potentials of a digitized public sector are regularly highlighted politically but have not been sufficiently exploited to date. In modern democracies, the public administration is part of the executive

Published by Springer Nature Switzerland AG 2022
R. Krimmer et al. (Eds.): ePart 2022, LNCS 13392, pp. 154–170, 2022.
https://doi.org/10.1007/978-3-031-23213-8_10

branch. These public authorities can roughly be divided into two execution levels. The first level depicts those authorities being mainly involved in the **enactment** of laws, e.g., ministries. Ministries are, for instance, involved in the legislative process by supporting the preparation or organizing consultations although not being part of the legislative as such. The second level depicts those authorities being mainly involved in the **enforcement** of laws, e.g., tax- or financial authorities. In this study, the former will be referred to as the enactment-level, the latter as the enforcement-level. With regard to public sector digitization, many scholars have focused on the investigation of the enforcement-level, e.g., by examining IT-related issues or scrutinizing organizational aspects [8,15]. Whereas these studies are undoubted of clinical importance, they leave an important factor out of account: the prerequisite to the work of enforcement-level, namely the laws they have to put into practice. These authorities are bound to these laws' content, whether they imply processes, data exchange, documentation, or reporting requirements. Consequently, the enforcement-level can only digitize, as the laws to be enforced allow it. Laws that are not digitally suitable make subsequent executive action more difficult. They hinder digital administrative processes and automated checks. Adherence to written form requirements slows down data collection and makes inter-agency data exchange difficult. Thus, the digital transformation will not succeed if the laws underlying the tasks of the enforcement-level are still designed for analog execution [22].

The described issue is not new to research and practice. On the one hand, some countries, first and foremost Denmark [5], have introduced measures for more digital-ready legislation. On the other hand, scholars have investigated how to improve the legislative process. A common approach is suggesting a so-called digitization check [21], which proposes to have draft legislation undergo an audit to evaluate its digital suitability. However, a shortcoming of current research is its underlying silo thinking: it only regards the enactment-level and respective processes. The role of the enforcement-level in the legislative process and the communicative aspects between both levels have not been regarded at all. The design of the new legislation is a complex process with many actors involved. It, therefore, does not suffice only to regard the enactment-level, but requires considering the enforcement-level and the interaction between both levels during the process. Based on the above, the research objective (RO) of this study is:

RO: The establishment of a set of design requirements for an institutionalized digitization check. The requirements aim at considering both levels of public authorities and their communicative intersection.

The remainder of this paper is structured as follows. In the next chapter, we present the research background to the content of this study. Afterward, we elaborate on the research design to increase the comprehensibility of how we reached our results. The findings are introduced before we derive the digitization check's design requirements and content-related aspects. These findings are discussed with respect to current examples from research and practice before we conclude with the limitations of our work and suggestions for future work.

The findings offer a substantial contribution to government digitization, as they bridge the gap between enactment- and enforcement-level. Moreover, it provides distinct action options for practitioners involved in improving the process toward digital-ready legislation.

2 Research Background

Administrative digitization is heavily dependent on legal regulations. Unlike in the private sector, legal regulations define communication channels and formal requirements weigh upon public authorities [11]. This dependency implies high demands on the design of laws relating to administrative services to promote the expansion of e-Government. To date, however, much research has focused on the legal situation being a barrier to implementation [23], as digital means are designed in a too complicated manner. Sometimes, such means are not even provided for in the first place, as in the case of mandatory written form requirements [17]. In order to address this problem directly at the initial stage, i.e., during the legislative process, digital enforceability of laws has meanwhile gained significant interest in science and practice. A central aspect of the often lacking digital enforceability goes back to the many actors involved, each with different interests. The incentives at the political level to drive legislation forward are different from those at the implementation level [22,25]. It may thus be opportune for political actors to pass laws as quickly as possible to gain political capital out of them. This, in turn, may lead to insufficient consideration of digitization aspects.

A kind of "digitization check" is increasingly being called for as a countermeasure. Such a check is understood in the following as an examination of draft laws for their digital enforceability, feasibility, and digital meaningfulness. Based on these discussions, some institutions have already started implementing corresponding mechanisms in the legislative process. The EU, for example, set up guidelines for a more sophisticated digital implementation of laws, which are laid out in *the Better Regulation Toolbox* [9]. The idea of such a check is not entirely new. Over the past years, several scholars have been investigating the nature of digital legislation from different perspectives. Hildebrandt, for example, vouched for legislators to revise the grammar and alphabet of laws to fit the requirements of a data-driven society [14]. Wong introduced the *seven levels of digitisation*, referring to write rules (=laws) as code [28]. The legislative process and its relation to modern automated decision-making have extensively been discussed by Schartum [24]. Novak et al. made suggestions on potential steps to revise the legislative process, focusing on international examples [21].

One of the very few examples in which practitioners have followed such suggestions is Denmark [5]. The country is considered a pioneering nation in introducing so-called digital-ready legislation, which relies on digital monitoring throughout the entire legislative process. This is based on seven principles: (1) simple, clear rules, (2) digital communication, (3) possibility of automated case processing, (4) consistency across authorities—uniform concepts and reuse of data, (5) safe and secure data handling, (6) use of public infrastructure and (7) prevention of fraud and errors. By putting these principles into practice, Denmark has increased its public value creation and thus has become a role model for the digital transformation of public administration [26].

However, even though examples like Denmark are essential, one cannot force the same system to work in other countries without further ado. It makes a significant difference, which governmental system a country has employed [1,13]. In Europe alone, seven different governmental systems are in place, e.g., the parliamentary system, the semi-presidential system, or the parliamentary monarchy [19]. Furthermore, even if countries belong to the same category, they are only hardly comparable. The authoritative organization of a single system can yet again have several instances: Germany, for example, is organized as a federal republic, meaning that there are different governmental levels. Denmark, in contrast, is organized centrally, which makes a massive difference when pursuing digitalization efforts [2].

To sum up, many scholars have already started investigating how to enable digital transformation through more digital-ready legislation but it has not grown to a mature field of research within the e-government domain. For instance, a holistic view of the main weaknesses in the legislative process has not yet been given. Too often, scholars have not considered the interests and needs of both the enactment and enforcement level. Thus, recommendations for action are scarce.

3 Research Design

To meet our research objective, we first analyzed the assembly of existing digitization checks to set up the interview study afterward. This analysis focused on existing approaches to digitization checks, such as Denmark's *seven principles for digital-ready legislation*[1], the *digital check* of the German federal state of North Rhine-Westphalia[2], the New Zealand *Better Rules for Government Report*[3], the *CIO Check* of the e-Government Act of the German federal state of Saxony[4], and the *better regulation guidelines and toolbox* published by the European Commission[5]. The individual facets of these checks were examined with a particular focus on the actors involved, the content of digitization checks, the binding nature, and the specific objectives.

The findings obtained in the first phase were then used to gain a better domain understanding. This was required to design the interview guide and determine which experts should be considered for the interview study. To this end, an interview guide was developed that enabled such a digitization check to be considered from four angles:

[1] Denmark: Seven principles for digital-ready legislation.
[2] North Rhine-Westphalia: Digitalcheck.
[3] New Zealand: Better Rules For Government Report.
[4] Saxony: e-Government Act 17(2).
[5] European Union: Better Regulation Guidelines and Toolbox.

1. the initial situation and existing challenges for a digital-ready legislation
2. the shortcomings of the legislative process on the enactment-level
3. the shortcomings of the legislative process on the enforcement-level
4. the shortcomings regarding communication aspects between both levels

In sum, we conducted nine interviews with experts from the field. Seven of them were experts from the domain under investigation—public officials working in the area of pension schemes. Those interviewees were involved in the same legislative proposal, that gained much attention and critique for its (digital) non-feasibility in 2019 and 2020 [10]. In addition, they were the ones tasked with implementing the aforementioned law having the following positions: top management, leading project management, judicial office, head of IT-department, head of the digitization-department, and two experts from the field of IT. The other two experts are working on establishing digitization checks, one interviewee on the federal and the other interviewee on the states level.

We performed the expert interviews according to a well-established method for qualitative interviews in IS research [20]. The expert interviews took 52 min on average, ranging between 25 and 69 min. All interviews were recorded and then transcribed. The subsequent coding and analysis process was organized as follows. To ensure having expertise from the relevant fields at hand—law, digital government, and administrative organization—we worked in an interdisciplinary team of five researchers with a background in e-government, information management, and law. The angles mentioned above served as a basis for the coding, and two persons coded every interview. After comparing the codes and adjusting them, they were clustered. The structured results were then synthesized, contextualized, and prepared for further processing [18]. For example, starting from the angle "initial situation and existing challenges", a sub-cluster "lack of digital skills" was derived, entailing the codes "age" and "educational issues."

All interviewees were German and therefore discussed the German perspective on the legislative process. Being fully aware of the potential problems of the country specificity, we decided to formulate the design requirements on an abstract level to detach the generated knowledge from its original national context and allow practitioners to adapt the requirements to their individual needs.

4 Results

The following paragraphs contain the interview results, the design requirements, and the content related aspects of the digitization check.

4.1 Existing Challenges for Digital-Ready Legislation

The introduction of a digitization check is not an entirely new topic. However, previous deliberations have not yet led to any concrete results. Based on the interviews conducted, there are four dimensions in which environmental challenges are reflected: (1) structural, (2) political, (3) personnel, and (4) technical.

At this point, it should be noted that this study has two research foci: an institutionalized digitization check at the enactment-level and an internal digitization check at the enforcement-level. The following challenges concern both levels.

Structural: Much coordination is required due to the variety of actors involved in the legislation and the operationalization of laws. Moreover, the secrecy status of relevant documents sometimes depicts a problem, as it may hinder key players from partaking in discussions due to the lack of necessary clearances.

> *"Sometimes we get something before the draft law, but it has such a secrecy status that it is only considered in the professional department [...]. However, that would be the time period at which the digitization department would also already have to be included."*

Political: The legislative process is often constitutionally not specified and has thus no fixed time-frame to complete. This is a problem for the implementers (=enforcement-level) because projects often have to be completed under high time pressure due to the political will to implement them. This impediment is additionally fueled by "thinking in terms of legislative periods", since there are often lengthy negotiations, leaving little time to ensure smooth implementation.

> *"The problem is that negotiations often take a very long time until the 'whether' of the law has been established, and then it's a question of the 'how'. And for the 'how' there is often not much time. By the time we get there, the law is already fully drafted, because the legislative period is over, and everything must be completed."*

Furthermore, laws are designed based on case-by-case justice to cover all conceivable scenarios, which presents digitization and IT experts with major hurdles since information systems should be based on standard variants; special cases should be exceptions.

In addition, public authorities of all levels have to deal with questions of departmental sovereignty and power shifts: new, digitized processes usually also bring new actors with them. This often meets resistance from parties whose decision-making-powers are questioned. Another circumstance describes the political self-rationality of actors in legislation. Often, politically desired laws are passed "as quietly as possible", taking into account the interests of the respective parties in power. Their priority is to cover qualitative aspects of a law, which often does not include digital enforceability. This self-rationality is characterized by the political landscape's growing competitive pressure between the actors.

> *"I do have the feeling that a lot of our arguments are being heard and dealt with, but sometimes there are certain political positions that are being pushed through."*

Personnel: Digitization projects are usually associated with organizational adjustments. Problems can arise if previously involved units do not feel sufficiently taken into account or people who are averse to digitization projects are in powerful positions. Moreover, the shortage of skilled workers, the difficulties in training staff, and the *fight for talent* depict challenges. Furthermore, the interviewees described a poorly developed digital mindset among existing employees. Laws, processes, and structures conducive to the digital implementation of legislative projects are still very much thought of in analog patterns, and therefore digitization potential cannot be tapped. Reasons for that are the workforce's increasing age and the inadequate training of existing staff.

> *"It certainly plays a role that digitization expertise [...] is not yet really available."*

Technical: Public authorities' historically evolved and heterogeneous IT and IS landscape further complicate all circumstances above. The diversity of the various data input channels (e-mail, mail, fax, etc.) and the associated patchy data availability were cited as the most significant challenges for the digital implementation of laws, as common information systems cannot process unstructured, incomplete data. This challenge is mainly reflected when technical projects are anchored in legal texts, associated with considerable implementation effort with a short time horizon. The matter becomes particularly problematic when a respective law implies the development of new information systems.

4.2 Shortcomings of the Legislative Process on the Enactment-Level

Interviewees have emphasized difficulties for effective consultation of external stakeholders in preparing legislation and the subsequent parliamentary legislative process. They mainly have criticized the short time frames for consideration of draft legislation given the complexity of laws. Not rarely are affected parties only given a couple of days for revising the proposition. This is particularly the case for legislative processes involving digital enforcement because proposed laws must not only be tested for their content but for their technical feasibility.

> *"[...] but three to four days? For an authority of our size, where we have hierarchical levels and many departments affected? We have to write a statement, that still has to be coordinated with the board of directors. Five days are a joke to be able to evaluate laws carefully."*

A major drawback is that stakeholders' comments outside the ministerial administration are currently not sufficiently audible. Expert hearings in parliament or committees are often a formality, not a serious attempt to improve the legal text. Given political self-rationalities, the legislator often aims to implement laws at the expense of appropriate participation periods rapidly. Additionally, many experts expressed the need for more process-oriented thinking as a prerequisite for the digital suitability of laws. Too often, processes that will be affected by a specific law are not appropriately modeled or formalized, so it becomes difficult for involved parties to have a discussion upon this basis.

"You have to have knowledge of new technologies [...], you have to look more at what you can do there. But what is always very important to me is to understand the technical processes, because often problems are not only technical topics, but also process topics."

Complementing the previous point, legislators often lack a thorough understanding of existing data and data flow within and between authorities. This, however, is also of central importance for the digital enforceability of laws. Without a clear vision of existing data and potential requirements of new infrastructure, a profound picture of a law's impact can hardly be drawn.

"A digitization check should look at who has what data and when, and then also look very closely at time periods, people and references."

The lack of mandatory legal regulation for a digital check is a final consideration. Without committal obligation, compliance to any procedure or check is unlikely to succeed.

4.3 Shortcomings of the Legislative Process on the Enforcement-Level

The following results depict obstacles, that the interviewees deemed especially relevant to the enforcement-level, certainly also due to their operational positions in implementing enacted laws. However, the reverse may not be to regard these problems as exclusive to this level; it is well possible for similar issues to arise at the enactment-level as well. Due to the enforcement-level focus of this study, however, this does not become evident in our results.

Legislation is often considered a matter of only the enactment-level, but the enforcement-level plays a crucial role in implementing enacted laws. It, therefore, makes sense to go beyond the actual legislator and expand the digitization check to the enforcement-level. An authority-internal digitization check would ensure that necessary technical expertise is incorporated into the statements towards the enactment-level and politicians. Organizational challenges refer to general prerequisites necessary for the efficient instantiation of a digitization check at the enforcement-level; specifically, interviewees were most concerned with an internally coordinated process and uniform procedures.

Experts repeatedly emphasized the lack of a coordinating office in the authority that bundles all participation requests and acts as a uniform office facing the legislator. Such a coordinating body may be the interface between the enactment-level (institutional) and the enforcement-level (internal).

"I see a key point in the question of how cross-cutting issues such as digitization can be really well coordinated."

Currently, points of contact are unclear for the enactment-level when proposing new legislation, especially when many actors are involved within an authority on the enforcement-level. An internal coordinating office could be the single point of contact and the uniform and coordinated appearance in participation procedures. As draft legislation typically concerns more than one department, coordination and a consistent consolidation of the positions are required. Interviewees highlighted the risk of potentially inconsistent statements from different organizational areas within the authority. An enforcing factor is the sensitivity that the communication in the "political negotiation process" demands.

"As long as I can't consolidate that [answers to the legislator] and give the answers with a big overview, then answers of individuals turn out differently and bind with it others, retrospectively. [...] Then there is a statement that is in the world. And in a political negotiation process, it is sometimes very difficult to come down from such statements."

In addition to organizational aspects, general digitization factors indirectly affecting a digitization check also were essential for the interviewees. For example, much of the present culture in authorities of the enforcement-level seems not to be very open towards digitization and the acquisition of more digital competence. This, however, is critical as the level of digital competence influences the quality of statements towards proposed laws regarding their digital enforceability.

4.4 Communication Issues Between Both Levels

A key finding of the interviews is that the communicative exchange between both levels does not follow a standardized process. Accordingly, it is not specified whether, when, and how communication occurs. This, however, must be ensured to establish the link between a digitization check at either of the levels. Communication or cooperation on the draft law before its release is rare. It is usually unclear when the enforcement-level will learn about the contents and technical requirements. However, it is essential to get involved early to consider IT-design and digitization. Communication is partly dependent on personal relationships. The interviewees reported that it was hardly ever possible to inspect preliminary drafts, even if they had good contacts; and if it was the case, the secrecy status of these drafts made it impossible to use further the information obtained.

"And it is often the case that, even with good contacts, you can hardly get hold of actual preliminary drafts [...] that you would need in order to already think in terms of technical design."

This was observed mainly in the case of politically controversial topics. Indeed, there may still be some changes in content before the draft bill is published. However, the risk of hasty implementation work was considered low, whereas the negative consequences of too late involvement were seen as severe.

4.5 Design Requirements

After the variety of impediments to successful digital-ready legislation had been discussed, the interviewees were also asked for potential solution approaches. From a multitude of different answers, a set of design requirements has been synthesized. Therefore, the first result of this study is a set of requirements, which is depicted in Table 1. The requirements can either be applied to the enactment-level (Enact.), the enforcement-level (Enfrc.), or both. The latter primarily concerns communication-aspects between the two levels.

Table 1. Design requirements for a digitization check

Design requirement	Enact.	Enfrc.
Introduction of a centralized digitization check	x	
Creation of a strong legal mandate	x	
Provision of the check with profound institutional anchoring	x	
Commitment of the parliament to the results of the check	x	
Standardization of draft legislation	x	
Early implementation with reference to the enforcement level	x	x
Establishment of systematic change management	x	x
Enforced competence building at all levels	x	x
Harmonization of IT landscapes	x	x
Introduction of innovative forms of exchange and collaboration	x	x
Standardization of communication channels at the institutional level	x	x
Standardization of internal communication		x
Creation and continuous expansion of an organization-specific checklist		x
Establishment of a sophisticated knowledge management		x

Enact. = enactment-level; Enfrc. = enforcement-level

Introduction of a centralized digitization check: To improve digital enforceability, laws should be subjected to a digitization check. The institutionalization of the check is recommended at the highest governmental level.

Creation of a strong legal mandate: A binding legal regulation is necessary. Noncommittal rules do not seem to be sufficient; they are not being adhered to.

Provision of the check with profound institutional anchoring: The digitization check should be assigned to a body that is, if possible, already entrusted with reviewing draft legislation. It should be an independent body with the (professional) prerequisites for a thorough digitization check.

"I would like to see a coordinating body that would bring together the results again and again and sharpen the view again and again and thus provide a rough direction."

Commitment of the parliament to the results of the check: If laws originate from within parliament (not directly from the government), the performance of a digitization check must also be guaranteed. The body performing the check must be able to exercise its mandate in this case, too. Parliament should commit itself to forwarding submitted drafts to the review body without delay.

Standardization of draft legislation: It is of great advantage for technology-supported feasibility of laws if they are based on standard variants and individual cases are exceptions. The aforementioned principles of the Danish "digital-ready legislation" can be seen as an example here.

"*Clear rules, clear terminology, avoid vague legal terms.*"

Early implementation with reference to the enforcement Level: The results show that it is essential to carry out the digitization check as early as possible. The institution performing the check should be involved in the preparation of the legislation at the same time as bodies from the enactment-level.

Establishment of systematic change management: Resistance to change [6] within organizations must be counteracted. The appropriate methods should be applied consistently and systematically in order to make the benefits and motivations of organizational changes clear to all those involved.

Enforced competence building at all levels: Public authorities face the effects of demographic change. To promote a digital mindset and ensure that digitization becomes a cross-cutting issue, the training of digital skills must be promoted and expanded outside of formal educational contexts.

Harmonization of IT landscapes: Numerous systems, patchy data availability, and a variation of digital and analog data input channels make it increasingly difficult to maintain requests digitally. For this reason, the system landscape required for a law and the required data input should be systematically investigated and tested so that digitally designed laws can also be implemented technically.

Introduction of innovative forms of exchange and collaboration: Innovative forms of cooperation, for example, in the form of "digital-legislature labs", should be used for suitable projects. These labs could serve to approach various topics at an early stage of the preparation and involve the enforcement level to test the requirements for digital suitability. As labs might not be appropriate for every law, they can be an optional supplement to the digitization check.

> "*We need to try out other formats of cooperation. In one example, workshops were held in which user journeys were created, which was a new experience for those involved. However, it was possible to see at an early stage what effects this would have on the implementation of this law. Even down to the affected person.*"

Standardization of communication at the institutional level: Relevant actors to the process must be clearly documented. This includes the creation of a coordinating body within the authority of the enforcement-level that bundles inquiries in the participation process to consolidate internal opinions from various specialist areas. This will ensure consistent and coordinated communication within the authority and the enactment-level.

Standardization of internal communication: When an authority is requested to submit a statement, a standardized procedure must be initiated. This procedure should consider all relevant actors as well as their tasks. It is imperative here that the review is carried out in an interdisciplinary manner to consider both the professional departments and the cross-cutting issue of digitization.

Create and continuously expand organization-specific checklist: The digitization check should be based on uniform, standardized criteria. The criteria collected in this study should be observed, supplemented by organization-specific criteria, and continuously expanded.

Implement knowledge management: To prevent knowledge-islands, measures should be taken to promote the organization-wide transfer of knowledge between individuals actively. This leads to institutional knowledge required to evaluate draft legislation not to get lost in case of retirement or other reasons employees would leave their job.

> "*As it is really important, we have actually set up a program with which we try to create know-how transfer, although it does not really succeed.*"

4.6 Contents of the Digitization Check

The previously presented conceptual design requirements depict the frame of a digitization check. Furthermore, the interviewees also attempted to indicate the content-related requirements of such a check. These items are not to be regarded as extensive nor exhaustive. They rather represent the most severe pain points that, from the interviewees' experience, have repeatedly played a role in legislative processes and impeded a proper execution. They thus can serve as a starting point for practitioners to draft the digitization check suitable for their context. The subjects are presented in Table 2, again differentiating between the enactment-level (Enact.) and the enforcement-level (Enfrc.).

Review possibility to automated data retrieval from existing systems: Does the draft law allow for automated retrieval and processing of the required data from existing IT systems?

Table 2. Subjects of the digitization check

Subject of dgitization check	Enact.	Enfrc.
Review possibility of automated data retrieval from existing systems	x	
Review possibility to use digital alternatives	x	
Modularize central legal concepts in draft laws	x	
Review possibility to make use of "once only"-data collection	x	
Carry out process modeling as part of the preparation process	x	
Examine affected actors and interdependencies		x
Investigate required data and affected data sets		x
Review of existing IT-systems for reusability		x
Review exisiting processes for adaptability and scalability		x
Review automation potentials of existing/planned systems		x
Review of required interfaces/data exchange platforms		x
Examination of the effort of digital implementation		x

Enact. = enactment-level; Enfrc. = enforcement-level

Review possibility to use digital alternatives: Does the draft law require written form requirements that could potentially be replaced with digital alternatives, such as electronic signatures?

Modularize central legal concepts in draft laws: Does the draft law allow for automated, rule-based processing, or is it built on case-by-case justice?

Review possibility to make use of "once only"-data collection: Does the draft law require the collection of data potentially already existing? Can the existing data be reused instead?

Carry out process modeling as part of the preparation process: Are the affected processes properly documented and well-understood by the affected parties? Is the law written in a way that these processes can accommodate it?

Examine affected actors and interdependencies: Which actors from the organization and the organizational environment are affected by the proposed legislation? Who must be involved?

Investigate required data and affected data sets: Are data required to implement the proposed law (e.g., for checking eligibility requirements) already available, or can they be made available?

Review of existing IT-systems for reusability: Which existing systems can be adopted or adapted?

Review existing IT-systems for adaptability and scalability: Is the processing of larger data volumes secured? Do our systems meet new requirements?

Review automation potentials of existing/planned IT-systems: To what extent can the processes resulting from the draft law be automated? Which effort is connected with adapting current systems/implementing new systems?

Review of required interfaces/data exchange platforms: Which interfaces between IT systems and/or organizations result from the proposed legislation? Which exchange channels can be used or must be created?

Examination of the effort of digital implementation: How much (financial and personnel) effort does it take to implement new processes digitally? What are the advantages compared to the implementation effort?

5 Discussion

The inhere presented design requirements for a digitization check address aspects of the enactment-level and the enforcement-level alike. Our work adds to the body of knowledge of the field, as it takes both perspectives into account. In previous works, the consideration has only occurred mono-perspectival. [14] took a perspective on the formulation of law. Specifically, the authors question whether the grammar and the alphabet of how the law is written meet the need of the data-driven society. They vouch for an understanding of *law as information* and therefore make it more accessible to computer systems. Building upon that, [24] investigates the relationship between traditional legislative systems and processes and their interference and interdependencies with governmental decision-making systems, considering computer-supported rule-based legislation. [22] look at the topic from a rather sociotechnical perspective, in which the digital-ready legislation is evaluated towards its consequences for the digitization efforts of public organizations. They conclude for digital-ready legislation to heavily influence the way of working of employees and the way the administration appears to the citizen. The authors also highlight potential downsides of such a legislative model; for example, decisions appear to be made randomly, and staff is reluctant to take responsibility for automated decisions. Finally, [21] have taken a rather practical stand on digital-ready legislation and propose to implement a digitization check, which is, however, not further specified.

Reconsidering the practical example given earlier, Denmark [5], we can see that the seven principles of the Danish digital-ready legislation are partly reflected in the Design Requirements provided in this article. Our Design Requirements mainly focus on the processual part of the legislation, i.e., *how to get to a law* that provides for digital applicability. In contrast, the Danish principles focus on the "endstate" of laws, i.e., *which characteristics* laws should have to provide for digital applicability. The design requirements and principles should thus be regarded as complementary. Both the design requirements and the seven principles sound universal at first glance, but they both impart critical aspects for digital legislation. As a next step, we must set forth to understand how to apply the rather universally formulated design requirements of this work to specific country contexts to enable governments to design digitally suitable legislation. While prior studies primarily regarded the enactment level of legislation (i.e., the legislative branch, not the executive), our article adds the enforcement-level perspective. This does not question the dominant role of the enactment level in the legislative process but calls for more thorough consideration of factors beyond the legislative branch, i.e., those actors enacting the laws once passed by the legislator. Our article provides a perspective match between both levels. Mutual understanding is crucial to building *common ground* [4] between two authoritative levels, working together to design laws that at the same time meet the requirements of a digitized society and still aim at providing the best outcome for a nation's citizens. As pointed out in the second chapter, considering the vast amount of different legislative and administrative systems and organizational architectures, implementing such digitization checks is surely not trivial. However, the design requirements of this study build upon an in democracies commonly used principle of division of powers between a legislative branch, an executive branch and a judicial branch. This principle is found on supra-national levels (e.g., EU), national levels (e.g., Germany) and sub-national levels (e.g., states). As long as the division of power is granted, the suitability of these design requirements is theoretically given for different forms of governmental organization (i.e., central or decentral).

6 Conclusion

The study shows that the existing potential of digitized administrative activities can only be realized if this goal is already taken into account in the preparation of legislation. The suitability of legal norms for execution depends largely on their digital enforceability. Laws suitable for digitization make it easier to implement the normative command and thus improve the law's effectiveness. Our findings show that adequate involvement of the enforcement-level and the authorities' interests is currently not ensured in the legislative process. The proposed design requirements contribute to better structuring of legislative preparation and more rational consideration of processes.

Despite conducting the research as rigorous as possible, it does not come without limitations. The experts interviewed were all of German nationality and thus concerned with legislation for Germany, which is why the recommendations may not be fully applicable to other countries. Even though we believe that the requirements established in this study are valid for those countries with a similar legal system as Germany (intertwined enactment and enforcement level), this assumption must be validated in the future. Furthermore, to have the majority of interviewed experts working on the same project offers both advantages and disadvantages. While this setting allowed us to examine a single project from a variety of perspectives relevant to digitization, it prevented us from confirming our findings with other projects. Thus, we cannot claim for our results to be exhaustive, as this would require the validation of our results on a broader scale.

Now that requirements for and subjects of a potential digitization check have been provided, scholars should attempt to understand the content-related dimensions of the digitization check better. Moreover, attention should be paid to the effects of the digitization check and whether it really improves laws.

To conclude, we can note the considerable potential for digital-ready legislation. This study contributes to better understanding the impediments that authorities from the enactment- and the enforcement-level face, and it provides practitioners with specific requirements they should adhere to when designing laws. After all, a more digital-ready legislation would benefit all parties: the enactment- and the enforcement-level as well as the citizenship and companies.

References

1. Boyne, G.: Local government structure and performance: lessons from America? Public Adm. **70**(3), 333–357 (1992)
2. Carstens, N.: Digitalisation Labs: A New Arena for Policy Design in German Multilevel Governance. German Politics (2021)
3. Choudrie, J., Weerakkody, V., Jones, S.: Realising e-government in the UK: rural and urban challenges. J. Enterp. Inf. Manag. **18**(5), 568–585 (2005)
4. Clark, H., Marshall, C.: Definite reference and mutual knowledge. In: Psycholinguisti. Critical Concepts in Psychology, 414P. Taylor & Francis (2002)
5. Danish Ministry of Finance; Agency for Digitisation: Agreement on digital-ready legislation. Tech. rep, Danish Ministry of Finance (2018)
6. Del Val, M.P., Fuentes, C.M.: Resistance to change: a literature review and empirical study. Manag. Decis. **41**(2), 148–155 (2003)
7. Dwivedi, Y.K., Weerakkody, V., Janssen, M.: Moving towards maturity: challenges to successful e-government implementation and diffusion. ACM SIGMIS Database DATABASE Adv. Inf. Syst. **42**(4), 11–22 (2012)
8. Ebrahim, Z., Irani, Z.: E-government adoption: architecture and barriers. Bus. Process. Manag. J. **11**(5), 589–611 (2005)
9. European Commission: Better Regulation Toolbox. Tech. rep, European Commission (2021)
10. Fachinger, U.: Plans for a basic pension in Germany revisited. ESPN Flash Rep. 2019/34 (2019)
11. Flechsig, C., Anslinger, F., Lasch, R.: Robotic process automation in purchasing and supply management: a multiple case study on potentials, barriers, and implementation. J. Purch. Supply Manag. **28**(1), 100718 (2021)
12. Goh, J.M., Arenas, A.E.: IT value creation in public sector: how IT-enabled capabilities mitigate tradeoffs in public organisations. Eur. J. Inf. Syst. **29**(1), 25–43 (2020)
13. Heinelt, H., Hlepas, N.K.: Typologies of local government systems. In: The European Mayor, pp. 21–42. Springer, Verlag (2006). https://doi.org/10.1007/978-3-531-90005-6
14. Hildebrandt, M.: Law as information in the era of data-driven agency. Mod. Law Rev. **79**(1), 1–30 (2016)
15. Jakob, M., Krcmar, H.: Toward an IT-strategy approach for small and mid-sized municipalities in a federal system. In: Proceedings of the 18th European Conference on Digital Government (ECDG), pp. 102–110 (2018)

16. Katsonis, M., Botros, A.: Digital government: a primer and professional perspectives. Aust. J. Public Adm. **74**(1), 42–52 (2015)
17. Kühn, H.: Recht als Gestaltungsinstrument einfacher, digitaler Verwaltungsleistungen – Bessere Rechtsetzung als Voraussetzung vollzugs- und digitaltauglicher Gesetze. In: Seckelmann, M., Brunzel, M. (eds.) Handbuch Onlinezugangsgesetz, pp. 17–51. Springer, Heidelberg (2021). https://doi.org/10.1007/978-3-662-62395-4_2
18. Mayring, P., et al.: Qualitative content analysis. Compar. Qual. Res. **1**(2), 159–176 (2004)
19. Molkenthin, M.: Die politische Europakarte (2022). http://www.europakarte.org/europakarte-politisc
20. Myers, M.D., Newman, M.: The qualitative interview in IS research: examining the craft. Inf. Organ. **17**(1), 2–26 (2007)
21. Novak, A.S., Huber, V., Virkar, S.: Digital legislation: Quo vadis? In: 22nd Annual International Conference on Digital Government Research (DG.O 2021), pp. 515–521 (2021)
22. Plesner, U., Justesen, L.: The double darkness of digitalization: shaping digital-ready legislation to reshape the conditions for public-sector digitalization. Sci. Technol. Human Values **47**(1), 146–173 (2022)
23. Savoldelli, A., Codagnone, C., Misuraca, G.: Understanding the e-government paradox: learning from literature and practice on barriers to adoption. Gov. Inf. Q. **31**, S63–S71 (2014)
24. Schartum, D.W.: Law and algorithms in the public domain. Nordic J. Appl. Ethics **10**(1), 15–26 (2016)
25. Schuppan, T., Köhl, S., Off, T.: Vollzugsorientierte Gesetzgebung durch eine Vollzugssimulationsmaschine. Tech. rep., Nationales E-Government Kompetenzzentrum e.V. (2018)
26. Scupola, A., Mergel, I.: Co-production in digital transformation of public administration and public value creation: The case of Denmark. Gov. Inf. Q. **39**(1) (2022)
27. Vial, G.: Understanding digital transformation: a review and a research agenda. J. Strat. Inf. Syst. **28**(2), 118–144 (2019)
28. Wong, M.W.H.M.: Rules as code-Seven levels of digitisation. Research Collection School of Law (2020)

Digital Society

Administrative Burden in Digital Self-service: An Empirical Study About Citizens in Need of Financial Assistance

Ida Heggertveit[1(✉)], Ida Lindgren[2], Christian Østergaard Madsen[3], and Sara Hofmann[1]

[1] University of Agder, Kristiansand, Norway
ida.heggertveit@uia.no
[2] Linköping University, Linköping, Sweden
[3] The IT University of Copenhagen, Copenhagen, Denmark

Abstract. The aim of this paper is to investigate what challenges arise for vulnerable citizens when welfare service provision is digitalized. We analyze the challenges citizens experience in the application process using the theoretical concept of administrative burden., i.e., learning-, compliance-, and psychological costs imposed on the citizen by policy implementation. The financial assistance service provided by the Norwegian Labor and Welfare Administration (NAV) is our empirical example. Our results show that digitalizing financial assistance creates new administrative burdens for vulnerable citizens. While frontline workers offer important help to citizens in the application process, they can also impose additional burdens on the citizen. Our study contributes with empirically grounded insights on the administrative burdens related to digital self-service, which causes citizens to turn to frontline workers for support. We offer a theoretical contribution by linking digitalization and administrative burden.

Keywords: Digital self-service · Administrative burden · Executive functioning · Financial scarcity · Digital divide

1 Introduction

There is a current trend towards digital self-service as first choice for citizens' interaction with government [1] as digitalization can provide citizens with easier access to public welfare services. However, even when digital self-service is the primary channel, citizens still turn to traditional channels, e.g., call centers and physical contact centers [2–4]. In the dawn of e-government, government organizations foresaw that digital self-service would reduce the need for call centers and physical meetings [5, 6]. However, citizens still use traditional channels, especially when they experience problems [3].

Channel choice (CC) research [3, 4, 7–9] has examined citizens' choice of communication channels in public service encounters and identified influencing factors related to the nature of service and citizens' interaction needs. Furthermore, CC is also influenced

Published by Springer Nature Switzerland AG 2022
R. Krimmer et al. (Eds.): ePart 2022, LNCS 13392, pp. 173–187, 2022.
https://doi.org/10.1007/978-3-031-23213-8_11

by citizens' digital skills, where scholars have illustrated clear inequalities in different citizens' ability to use technology [10]; sometimes discussed under the 'digital divide' label [11]. Recently, the concept of *administrative burden* has been presented as a possible influence on citizens' CC [12]. Administrative burden is defined as *"the learning, psychological, and compliance costs that citizens experience in their interactions with government"* [13]. The underlying idea is that public service provision is associated with a certain amount of work. Depending on how the service interaction is designed, more or less work is distributed to the citizen who uses the service [12]. Studies show that citizens in most need of governmental support are also the most likely to experience administrative burdens when applying for public welfare services [14]. In this paper, we refer to this group of citizens as *vulnerable citizens.* We especially focus on vulnerable citizens in need of financial support, as financial scarcity is often associated with increased stress-levels, potentially reinforcing administrative burdens and inequalities [14]. Previous studies have shown that digital self-service can impose administrative burdens on citizens in need of financial support, by forcing citizens to conduct work previously performed by professional caseworkers [12]. Furthermore, automated systems can impose administrative burdens on citizens that are not covered by the automation and therefore must apply manually [15].

Many studies have focused on what effects digitalization of public welfare services have on interactions between government and citizens [1, 3, 16, 17] and how relationships between frontline workers, citizens, and technology are affected by citizens' use of digital self-service [3, 16]. However, we need more in-depth studies on how digitalization of welfare service provision affects vulnerable citizens' channel behavior. Scholars have called for studies on citizens' channel behavior related to transactions, where the service process includes filling out forms, filing applications, and exchanging money [3].

The aim of this paper is to investigate what challenges arise for vulnerable citizens when welfare service provision is digitalized. We ask: *What challenges do citizens in financial need experience during benefit application that causes them to contact frontline staff for help?* We analyze the challenges experienced in the service process by applying a diagnostic tool for assessing administrative burdens [13]. We use the financial assistance service provided by the Norwegian Labor and Welfare Administration (NAV) as our empirical example. We study the challenges arising for citizens applying for financial assistance by interviewing NAV frontline staff who support citizens in their interactions, and other topic experts from NAV. Financial assistance is a fitting area for our study for two main reasons. First, NAV promotes citizens to apply for this benefit through an online self-service application. Second, and more importantly, financial assistance is a welfare benefit aimed to secure a minimum income and improve living conditions for vulnerable citizens. We therefore highlight the *scarcity* that occurs when you have less then you feel you need [18] to illustrate the role of individual experiences of financial scarcity as a prerequisite for applying for the service. Often, citizens applying for financial assistance are also experiencing other types of scarcity as well (lack of job, lack of social network) and health issues, adding to their vulnerability. Financial scarcity can lead to reduced individual cognitive ability and mental accounting [18]. These stress factors can influence the ability to cope with digital self-service application processes. However, this study will focus on the effect of financial scarcity.

Our study contributes with empirically grounded insights on how vulnerable citizens experience burdens when they are referred to digital self-service as their main channel of interaction with public organizations. We also offer a theoretical contribution by linking digitalization and administrative burden. The consequence of these burdens are important to convey and understand in order to avoid that structural societal inequalities [15] are reinforced by digitalization of public service provision.

This paper is organized as follows: Section two presents the concept of administrative burden and the role of executive functioning. Section three presents our research approach and the diagnostic tool used for analyzing administrative burdens. In section four, we present our findings and illustrate the main burdens found in the study. In section five, we discuss our findings and finally, we conclude the main findings and suggest areas for future studies on this topic.

2 Administrative Burden and Executive Functioning

Administrative burden refers to *"an individual's experience of policy implementation as onerous"* [13]. It was initially used to analyze problems employees experience in implementing and performing policy tasks in the public sector [13]. Herd and Moynihan shifted the focus to various costs that *citizens* might face in interaction with the state [13, 14]. Recently, information systems scholars expanded the concept to the e-government area [12, 15]. Administrative burden [13] consists of three components (Table 1). First, *learning costs* relate to the time and effort the citizen must invest to learn about the service and establish eligibility, what conditions to satisfy, and how to access the service. Second, *compliance costs* relate to resources and work the citizen needs to put in to reach eligibility status, e.g., collecting documentation and responding to discretionary demands made by the administrators. Third, *psychological costs* relate to the negative psychological effects, e.g., stress, frustrations, losing power or autonomy and stigma, arising from applying and participating in an unpopular program.

Table 1. The components of administrative burden [13], p. 23

Type of cost	Description
Learning cost	Time and effort expected to learn about the program or service, ascertaining eligibility status, the nature of benefits, conditions that must be satisfied, and how to gain access
Compliance cost	Provision of information and documentation to demonstrate standing; financial costs to access services (such as fees, legal representation, travel costs); avoiding or responding to discretionary demands made by administrators
Psychological cost	Stigma arising from applying for and participating in an unpopular program; loss of autonomy that comes from intrusive administrative supervision; frustration at dealing with learning and compliance costs, unjust or unnecessary procedures; stresses that arise from uncertainty about whether a citizen can negotiate processes and compliance costs

Herd and Moynihan suggest that human capital e.g., money, social networks, intelligence, health and education, influence how people cope with and are affected by administrative burdens [13]. Together with Christensen et al. they argue that citizens' degree of executive functioning can explain why individuals are influenced differently by administrative burdens. They present a model of how citizens' executive functioning influences their ability to respond to state actions (see Fig. 1). Executive functioning is a neuropsychological construct that refers to an individual's ability to *"(1) reason and generate goals and plans, (2) maintain focus and motivation to follow through with goals and plans, and (3) flexibly alter goals and plans in response to changing contingencies"* [19]. These are essential abilities when identifying, understanding, and complying with public programs, services, and benefits. Citizens with low executive functioning often have difficulties in forming goals, planning, carrying out a plan and preforming effectively. Further, they may be challenged by short-sighted temptations, reduced emotional control, and poor social regulations [14]. Executive functioning is negatively affected by individuals' experience of scarcity (e.g., lacking money, time, or social support), health issues (e.g., mental problems, depression, physical pain, and the use of medications and drugs), and cognitive decline (e.g., age-related cognitive decline) [14]. Notably, even highly educated, and digitally skilled citizens may experience problems interacting with the government when they experience health issues. Moreover, citizens with low executive functioning are more likely to experience burdens in these interactions, and these burdens, in turn, are likely to reinforce citizens' problems by inducing stress.

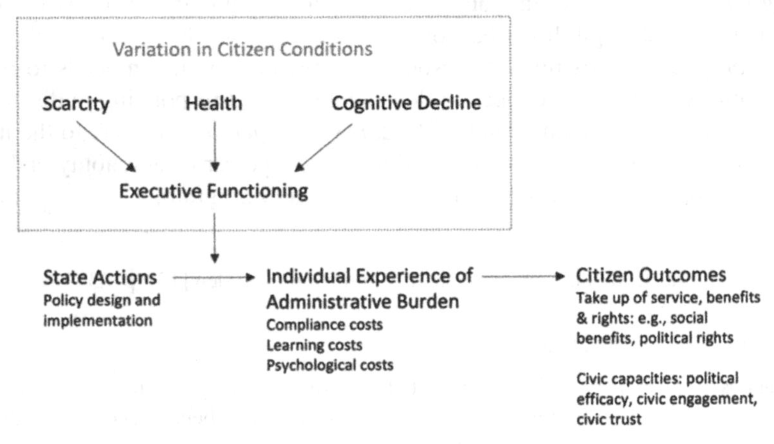

Fig. 1. Human capital and administrative burden: the role of executive functioning [14]

Scarcity refers to *"having less of something that you feel you need"* [18]. Our focus is on financial scarcity. Experiences of financial scarcity can explain why some people find a set of rules, forms and procedures more emotionally difficult and onerous than others [14]. We find the concept of administrative burden particularly fitting for a study of public services related to financial assistance, because citizens who apply for financial support often find themselves in complex situations and their experiences surrounding their financial situation tend to be stressful.

3 Research Settings

This study was conducted as part of a larger research project financed by the Norwegian Labor and Welfare Administration (NAV) [26]. Our work is qualitative and within the interpretive research tradition [20].

We focus on *financial assistance,* which is a welfare benefit offered to financially vulnerable citizens in Norway. Financial assistance and its eligibility criteria are based on the Act of Social Services. Financial assistance is a temporary benefit with the aims to cover citizens' necessary expenses. Rates are based on individual assessments, but they are often compared to national guidelines for living costs. It can be difficult for citizens to understand if they lack income to cover the basic because this is related to a subjective feelings of scarcity. It can also be difficult to understand what expenses are covered by the national guidelines. Everyone has the right to apply for this benefit and to have NAV conduct an individual assessment of their application. The purpose of this welfare benefit is to secure a minimum income, improve living conditions, promote transition to work, social inclusion, and active participation in society. The provision of financial assistance typically involves two government roles: *frontline workers* and *case workers.* Frontline workers are those the citizens can call when they have questions about the application process. Case workers, in turn, are the professionals that assess and decide whether the citizen is eligible for support. For citizens to receive financial assistance, they must be able to show that they lack financial assets to cover fundamental costs (e.g., rent, heat, food, basic clothing). There is no fixed rate, meaning that the financial assistance is based on the individual need of the citizen applying for support. NAV promotes citizens to apply for financial assistance by using a digital self-service to make the service more efficient. In addition, caseworkers can allocate time for job-oriented counseling [21].

4 Method

We conducted three in-depth semi-structured interviews with six employees at NAV. The purpose of these interviews was to identify the steps in the application process, the channels citizens use, and what problems arise during citizens' interaction with NAV. We interviewed NAV employees, rather than citizens, because employees are experts with aggregated knowledge of the entire service process we are studying. Through their work, they have gained an overview of the citizens' most frequently asked questions and encountered problems. We regard NAV employees as "highly knowledgeable informants", who provide a collective picture and useful retrospectives [22]. Frontline workers are experts in the form that they have privileged access to citizens and the decision-making service process. Interviewing experts can provide knowledge and orientation to the specific field. They provide high technical knowledge of the field in relation to information needed to apply and the application procedure. Furthermore, they provide us with knowledge about interactions, routines, and social practices [23].

All interviews were conducted and recorded via Microsoft Teams and transcribed.

To complement the interviews, we analyzed NAV's official website and read available information on financial assistance. We also studied the design and content of the digital form that citizens need to fill in to apply for financial assistance. Further, we studied internal training documentation for caseworkers on how to promote citizens to use digital channels. Lastly, we analyzed the results from NAV's yearly analysis of future challenges and trends (called *Horizon Scan*) and their user survey on citizens' chosen communication channels and typical problems in interacting with NAV. Table 2 provides an overview of our empirical data.

Table 2. Empirical data: An overview.

	Respondent	Time	Period
Interviews	Interview with two representatives from the digitalization group within NAV, including a UX designer (UXD) and a caseworker (CW)	1 h 17 min	Autumn 2021
	Interview with two frontline workers from NAV call center, (FW1 & FW2)	1 h 50 min	Autumn 2021
	Interview with two frontline workers from NAV call center, (FW3 & FW4)	1 h 15 min	Autumn 2021
Documents	Website of NAV Digital application form Internal training document for caseworkers NAV's Horizon Scan NAV's yearly user survey		Spring 2022

We analyzed the empirical material using a qualitative content analysis approach, as suggested by Mayring [24]. The analysis was both deductive and inductive. The analysis was deductive in the sense that we critically analyzed the material using the diagnostic questions for identifying administrative burdens, as presented by Herd and Moynihan (see Table 3). During the analysis, we also allowed for inductive themes to emerge from the material.

Table 3. Diagnostic questions about administrative burden [13] p. 258

Component	Diagnostic questions
Learning costs	Is it easy for potential participants to Find out about the program? Establish if they are eligible? Understand what benefits are provided? Learn about application processes?

(*continued*)

Table 3. (*continued*)

Component	Diagnostic questions
Compliance costs	How many questions and forms are there to complete? How much documentation is needed? Do the participants have to input the same information multiple times? Is the information sought already captured via administrative data? Is it possible to serve the person in a less intrusive way, such as phone rather than in-person interviews? Do applicants have easily accessible help? How frequent is reenrollment? How much time must people commit to the process? What are the bottlenecks? What are the financial costs?
Psychological costs	Are the interactions stressful? Do people receive respectful treatment? Do people enjoy some autonomy in the interaction?

5 Findings

This section presents our analysis and findings on what challenges arise for vulnerable citizens when they apply for financial assistance. Citizens contacting the call center have usually tried to help themselves through the self-service but have run into problems. When a citizen contacts NAV call center by chat or telephone, the frontline worker judges the citizen's cognitive and digital skills, e.g., language and general understanding, and access to Bank-ID (necessary to use the application). Citizens who are found to have sufficient skills, are referred to seek information themselves on the webpage. "*We very often refer to nav.no, where it is very well explained*" (interview, FW3). Citizens who are found to have insufficient understanding, poor language skills, or need physical assistance, are sometimes offered a physical meeting with a caseworker at a local NAV office. Our study does not account for what happens during physical meetings. Below, we account for those issues that are typically dealt with by frontline workers through telephone, chat, and email, e.g., whether the citizen is eligible for financial assistance, what they can apply for, what documentation they need to attach and how to fill out the form correctly. We structure our findings based on the overarching components of administrative burden; learning-, compliance-, and psychological costs.

5.1 Learning Costs

When in need of financial assistance, citizens must first learn about the service and how to apply. The frontline workers receive many questions from citizens that illustrate difficulties in learning about these aspects. We identified two salient themes related to learning costs in the application process: 1) difficulties in understanding eligibility criteria, and 2) language barriers and lack of bureaucratic skills.

5.1.1 Difficulties in Understanding Eligibility Criteria

There are several learning costs associated with understanding and determining the eligibility criteria for financial assistance for vulnerable citizens. According to the frontline workers, citizens typically contact them as a complement to navigating the website. Some eligible criteria are mentioned on the webpage, such as having a legal residence in Norway and not living abroad when applying. There is also a video about the application process on the NAV webpage on how to "get started with your digital application", but only in Norwegian. Still, citizens need to determine whether they are eligible and what they can apply for. For example, the webpage informs the citizens that they can apply for living expenses, using examples such as food, rent, electricity, and clothes. Even so, the citizens struggle to understand what to write in the self-service application's open text field *"what are you applying for"* and the additional field *"give a brief justification for the application*".

For some citizens, financial assistance is just one of several economic benefits they apply for. Financial assistance covers many aspects (e.g., living expenses, housing expenses and dental treatment) and is often seen as the most suitable for urgent financial needs. As explained by the caseworker "*In my experience, many users do not think about what they may be entitled to before they apply. They often seek several benefits in hope of getting money. Because they need money. It is uncertain whether they understand the content and what they are actually entitled to"* (interview, CW).

According to the Act of Social Service, all citizens have a right to apply for financial assistance and to receive individual assessment of their application. Citizens can take a chance and apply even if they do not fully understand the eligibility criteria. To hinder unnecessary case work, frontline workers have a gate-keeping job for caseworkers when it comes to help citizens determine eligibility for this welfare service. For example, citizens who contact the call center, are asked if they have savings to use, an expensive car to sell, or other possibilities to be self-sufficient. Nevertheless, according to the case worker, *"many people apply for financial assistance without being entitled to it"* (interview, CW). Citizens and caseworkers can have a different view on the importance of keeping a car as a part of their life expenses. This is an example of an individual's experience of scarcity that can be difficult to explain in the free text area of the online self-service application.

There are different criteria for sending documentation based on what the citizen is applying for and if the citizen has applied earlier. Citizens struggle with finding the right digital application (if they have more than one) to attach the documents, and they struggle to find the correct documents and upload them. This part of the digital self-service application causes many applicants to abort the entire process.

For emergency situations, case workers must respond within 24 h e.g., when the citizen is without food, or the electricity is shut down. The frontline worker often becomes an important gatekeeper and interpreter of the citizens' situation and must decide whether the situation is an emergency or not based on the citizen's accounts. According to the frontline workers, many citizens lack cognitive skills themselves to understand what is defined as an emergency. *"The user does not understand what an emergency is according to the Social Service Act"* (interview, FW4). *"Users that do not know what emergency aid means and think rent is emergency. They call and say they need an answer now, and they call several times a day. We must therefore handle these phones so that the caseworker is freed from spending time on 15–20 such phone calls during the day"* (interview, FW4).

5.1.2 Language Barriers and Lack of Bureaucratic Skills

Learning about NAV's benefits to citizens, and how to apply, is difficult for many citizens, vulnerable or not. In a survey performed by NAV, 35% of the respondents answered that it is difficult to understand information about rules and duties. About 40% found it difficult to understand what rules are used to make the decision and how NAV has arrived at the amount they have been paid [25]. Also, the terminology used on NAV's webpage is difficult for many citizens to understand. The frontline workers mentioned that many of the vulnerable citizens applying for financial assistance find it difficult to understand the definitions of the central terms used in the application process, such as "income". For example, some citizens do not know that their savings are defined as income.

In the self-service application, the steps in the digital application contain question marks next to them with additional information on each step. Still, many citizens fail to understand what is expected of them and *"they do not always have the skills to acquire that knowledge"* (interview, FW3). Citizens often find it difficult to understand the application form and terminology used in relation to their own life situation, which can be complicated and not necessarily match all the steps in the application form. Also, many citizens lack sufficient knowledge on how the public sector works, and *"young people in particular may not understand the bureaucratic language*" (interview, CW). According to NAV's own user surveys, young citizens with high education and good digital competence contact NAV more frequently to get answers to questions [25].

Sometimes, difficulties in understanding stem from insufficient language skills. For example, immigrants are overrepresented as applicants of financial assistance [25] and often do not speak Norwegian well enough to understand the information on NAV's webpage and self-service online. Citizens with dyslexia also have problems understanding the information in the digital channels. However, according to the frontline workers, burdens fall disproportionately harder on immigrants in contrast to citizens with reading and writing difficulties or dyslexia, where frontline workers can read the questions out loud, using a simpler language, when guiding them through the digital application.

5.2 Compliance Costs

The form used to apply for financial assistance has nine questions. In addition, applicants must supply NAV with documentation that illustrate their eligibility. Applying the diagnostic questions related to compliance costs in the process, we identified two interrelated themes; 1) difficulties in supplying the documentation needed, and 2) difficulties in getting informed help by frontline workers.

5.2.1 Difficulties in Supplying the Documentation Needed

Supplying NAV with the necessary documentation is difficult and this is the step in the application process where citizens spend the most time; and can hence be seen as a compliance cost. Many citizens also abort their application during this step due to problems with uploading the right documents. *"There is an incredible amount of documentation to attach (...) In a way, you must put together your entire accounts. I would think it's a full-time job for a few days to get everything in place"* (interview, FW2). This is particularly problematic the first time a citizen applies for financial assistance. If a citizen has applied for financial assistance before, it is possible to reuse previously submitted documentation to NAV.

The documents NAV require differ depending on why the citizen need financial assistance. Therefore, this requirement is not specified. Many citizens struggle with this and need help to find out. Frontline workers try to reduce this burden by recommending citizens to upload any document they see could help in assessing their eligibility. The motto among frontline workers is to advice citizens to upload too much, rather than too little. *"At least, that's my motto. Several of us who respond to financial assistance follow it"* (interview, FW4). This recommendation can create uncertainty and, in turn, become a burden if the citizen uploads unnecessary documents.

The frontline workers report that non-digital citizen face more compliance costs than citizens with digital competences. Non-digital citizens need to collect the application form at the local NAV office or print it out from the municipal's website (where this is available). If they need help in the process and go to a physical office to get help, they are limited by the opening hours. Also, they are limited by the fact that the frontline workers cannot access their personal information (as discussed below). Put together, *"they are the weakest in society, but they are also the weakest in terms of opportunities for us to look at their case and themselves to see"* (interview, FW3).

5.2.2 Difficulties in Getting Informed Help by Frontline Workers

When citizens need help, they are directed to the call center and its frontline workers. While frontline workers can help with many things, they do not have access to the citizens' personal files. The Social Service Act prevents frontline workers' access to the case and prevents frontline workers from sharing information and answering questions in written channels. *"I would like to emphasize that we generally have very little transparency. We cannot see what they have applied for, the documentation they have attached, or the decision letter. We provide guidance on a general basis and can only see small things as if the payment is on the way or not. Otherwise, we do not have access to the user's case"* (interview, FW3). Consequently, frontline workers cannot give informed help based on the specifics of the citizen's case, other than from what information the citizen is conveying during the telephone call. The frontline workers report that the amount of inquires solved in the frontline is low compared to other services: *"we pass on many of them (...) unfortunately, it is difficult when we have so little transparency. We can rarely answer them when they call"* (interview, FW1).

Citizens sometimes have personal case workers. Many citizens call claiming to have urgent needs, wanting to speak to their case worker. The relationship between the citizen and the case worker is created through physical meetings and phone calls. If the frontline worker considers a citizen's call and errand as urgent, they can transfer the call from the citizen directly to the caseworker. The caseworker can decline the call and tell the frontline worker to leave a message. *"Several users state that it was better to contact someone who could actually answer questions about their case because they had access to the professional system, instead of contacting someone who does not know anything about their case and then get more diffuse answers, or maybe not an answer at all. (…) I don't find it very surprising that users miss getting detailed answers. I've always thought it's been important"* (interview, CW). Frontline workers say that the best solution would be direct contact between the citizen and the caseworker, and that citizens should have more insight in their own cases. *"Because these are people in particularly vulnerable situations [it is important to] establish a bond between the user and caseworker to provide the best possible follow-up"* (interview, CW).

5.3 Psychological Cost

Citizens who need financial assistance are often marginalized (being immigrants, unemployed, suffering mental or physical illnesses, drug use, etc.). This can be associated with increased stress levels [14]. Our respondents state that that the benefit application process for financial assistance can add to this stress for vulnerable citizens. By applying the diagnostic questions, we identified one salient theme related to psychological costs: 1) Stress and frustration.

5.3.1 Stress and Frustration

According to our informants, citizens who apply for financial assistance express that they experience stress and frustration in the application process. This may be due to the problems discussed above related to learning and compliance. But stress and frustration can also arise for other reasons. The frontline workers express that most citizens who reach out for help in the application process say that they are under time pressure. Financial assistance is one of the services at NAV with the shortest lead time, from application to decision, but for many citizens in need of financial assistance, the lead time is not perceived as fast enough. From the frontline - and case workers' perspective, the feeling of time pressure is often a result of bad planning (on behalf of the citizen). Oftentimes, *"they somehow fail to envision case processing time and are in a constant backlog. It's always NAV's fault"* (interview, FW1). Many of the citizens applying for financial assistance are in a vulnerable situation and *"may not have the cognitive abilities to plan ahead"* (interview, FW1).

In general, although stressed, citizens usually manage to keep a good tone when they call the call center. In contrast, frustrated citizens often use a harder tone in written communication, e.g., in chat messages and emails ("write to us"-sites) where frontline workers are sometimes addressed as "idiots" and "stupid". These are clear signs of frustration and stress experienced by the citizens. Unfortunately, these outlets of frustration can negatively influence the citizen's opportunity to receive financial assistance by creating a feeling of unwillingness in the frontline worker: *"Today I had to ask someone in the chat to keep a friendly tone. I do not want to help people who aren't nice either. I will of course help, but not at any cost. We must be able to maintain a friendly tone"* (interview, FW1). The interactions are also influenced by the citizens disorders. *"My general impression is that there is a lot of intoxication and mental disorders overrepresented.*

These conversations are especially demanding because there can be shouting, yelling, insults and much more" (interview, FW4).

6 Discussion and Conclusions

The aim of this paper is to investigate what challenges arise for vulnerable citizens when welfare service provision is digitalized. We asked, *What challenges do citizens in financial need experience during benefit application that causes them to contact frontline staff for help*?

By referring citizens to help themselves through digital self-service, the citizens are forced to search the agency's webpage and self-service forms for information on what services they are eligible for, how to apply, and how to illustrate eligibility. Citizens in need of financial assistance are typically suffering from scarcity or health issues. Therefore, searching this information on their own can be difficult.

Our analysis, using the diagnostic tool for assessing administrative burden, shows that learning-, compliance - and psychological costs are induced in several ways (see Table 4).

Table 4. Summary of findings

Component	Findings
Learning costs	Difficulties in understanding eligibility criteria Langugage barriers and lack of bureaucratic skills
Compliance costs	Difficulties in getting informed help by frontline workers Difficulties in supplying the documentation needed
Psychological costs	Stress and frustration

Our findings are in line with previous research on administrative burden [13]. When digitalizing financial assistance, citizens must find out about the service on their own. They also need to establish how to apply and if they are eligible, based on information from NAV's webpage. This creates learning costs as the citizen must learn about both the benefit, and the specific vocabulary associated with it [13] Citizens with language barriers and low bureaucratic skills are overrepresented among those applying for financial assistance, making this burden harder [14]. When it is difficult to understand what is required, it is also difficult to comply with the service requirements [13]. We have identified several interaction effects between the three learning costs. For example, combinations of learning - and compliance costs are clearly illustrated in our case. The frontline workers testify that many of those who reach out to them for help find it difficult to understand NAV's information on what is required of them. Therefore, they also struggle to comply with the requirements. Herd and Moynihan [13] state that psychological cost can be an effect of learning - and compliance costs. In the empirical material we see signs of what Christensen et al. [14] illustrate, namely that low executive functioning negatively affects citizens ability to complete an application process, and that this in turn, reinforces stress levels and further reduces the citizen's ability to complete the process. Interestingly, we also illustrate how a citizen's stress and frustration can cause bad behavior towards the frontline worker, which in turn may reduce the frontline workers' willingness to help the citizen.

This study provides empirical examples of how administrative burdens can be manifested in e-government practice. By using the diagnostic tool for assessing administrative burdens [13], we provide clear examples of how digital self-serivce creates additional burdens for some of the most vulnerable citizens in society. We particularly show how already vulnerable citizens in society risk being affected by even more administrative burdens due to their life situation. Burdens hurt vulnerable people more than others, and therefore worsen existing inequality in society. This study provides attention to the consequences of digital developments in social welfare systems. Digital public service provision is often presented as a way of improving life and service quality for citizens. We show how digital self-service can do the opposite for a small, albeit important, group of citizens in society. For some citizens, it is vital to have personal meetings with professionals who can describe and explain the service requirements in simple terms and provide hands-on guidance through the application process.

7 Implications, Limitations, and Future Research

Our study contributes with empirically grounded insights on how vulnerable citizens experience administrative burdens when applying for financial assistance through digital self-service. These insights also provide theoretical contributions, linking digitalization and administrative burden. It is important to convey and understand the consequences of these burdens to avoid those structural societal inequalities that risk being reinforced by digitalization of public service provision.

Our study furthermore provides some practical implications, relevant to NAV. Our analysis shows that information about financial assistance and online self-service application makes it easier for citizens to apply for financial support from their own homes at any hour of the day. However, self-service requires that the applicants understand the eligibility criteria, how to fill out the form correctly and how to upload the right documents. Many citizens call to get information on the status of their application or try to influence the outcome of their case. Some even start new applications to get in contact with caseworkers at NAV while they are waiting for a reply on previously made applications. Citizens who have applied digitally can forward requested documentation and get an overview of their applications if they log in to NAVs webpage. However, citizens struggle with navigating their way to find this feature on the NAV webpage. Easier access directly between citizens and caseworkers, or more transparency in the application process for citizens applying, could help reduce psychological costs.

Our study has several limitations. First, it is only based on a few interviews. Second, there are no interviews with citizens included in the study; thus, we have no first-hand accounts of how citizens themselves experience the self-service process. However, our interviews with frontline workers and other topic experts have provided us with an aggregated view of the various problems citizens face. To expand the scope, the study builds on documentation where user surveys with citizens responds are involved. Nevertheless, future research is needed based on interviews with citizens and observations of citizens' behavior in the application process. We need further insights on individuals' variations and differences in using of self-services, their understanding of empowerment, and why some citizens chose to use digital self-services, and others chose to reject them. Therefore, the next step in our study is to expand our research of administrative burdens by involving the citizens applying for financial assistance.

7.1 Disclosure

The research for this paper was co-financed by the Norwegian Labour and Welfare Administration (NAV) and the Research Council of Norway [project number 316246]. First author has first-hand experience from working as a caseworker with financial assistance at NAV before the service became digital in the period 2013–2016.

References

1. Lindgren, I., Madsen, C.Ø., Hofmann, S., Melin, U.: Close encounters of the digital kind: a research agenda for the digitalization of public services. Gov. Inf. Q. **36**, 427–436 (2019)
2. Madsen, C.Ø., Hofmann, S., Pieterson, W.: Channel choice complications: exploring the multiplex nature of citizens' channel choices. In: Lindgren, I., et al. (eds.) Electronic Government: 18th IFIP WG 8.5 International Conference, EGOV 2019, San Benedetto Del Tronto, Italy, September 2–4, 2019, Proceedings, pp. 139–151. Springer, Cham (2019). https://doi.org/10.1007/978-3-030-27325-5_11
3. Pieterson, W.J., Ebbers, W.E.: Channel choice evolution: an empirical analysis of shifting channel behavior across demographics and tasks. Gov. Inf. Q. **37**, 101478 (2020)
4. Madsen, C.Ø., Kræmmergaard, P.: Channel choice: a literature review. In: Tambouris, E., et al. (eds.) Electronic Government: 14th IFIP WG 8.5 International Conference, EGOV 2015, Thessaloniki, Greece, August 30 -- September 2, 2015, Proceedings, pp. 3–18. Springer, Cham (2015). https://doi.org/10.1007/978-3-319-22479-4_1
5. Layne, K., Lee, J.: Developing fully functional E-government: a four stage model. Gov. Inf. Q. **18**, 122–136 (2001)
6. Norris, D.F., Reddick, C.G.: Local E-government in the United States: transformation or incremental change? Public Adm. Rev. **73**, 165–175 (2013)
7. Reddick, C.G., Anthopoulos, L.: Interactions with e-government, new digital media and traditional channel choices: citizen-initiated factors. Transforming Gov. People Process Policy **8**, 398–419 (2014)
8. Ebbers, W., Pieterson, W., Noordman, H.N.: Electronic government: rethinking channel management strategies. Gov. Inf. Q. **25**, 181–201 (2008)
9. Teerling, M.L., Pieterson, W.: Multichannel marketing: an experiment on guiding citizens to the electronic channels. Gov. Inf. Q. **27**, 98–107 (2010)
10. Ebbers, W.E., Jansen, M.G., van Deursen, A.J.: Impact of the digital divide on e-government: expanding from channel choice to channel usage. Gov. Inf. Q. **33**, 685–692 (2016)
11. Van Deursen, A., Van Dijk, J.: Internet skills and the digital divide. New Media Soc. **13**, 893–911 (2011)
12. Madsen, C.Ø., Lindgren, I., Melin, U.: The accidental caseworker–how digital self-service influences citizens' administrative burden. Gov. Inf. Q. **39**, 101653 (2022)
13. Herd, P., Moynihan, D.P.: Administrative Burden: Policymaking by Other Means. Russell Sage Foundation (2019)

14. Christensen, J., Aarøe, L., Baekgaard, M., Herd, P., Moynihan, D.P.: Human capital and administrative burden: the role of cognitive resources in citizen-state interactions. Public Adm. Rev. **80**, 127–136 (2020)
15. Larsson, K.K.: Digitization or equality: when government automation covers some, but not all citizens. Gov. Inf. Q. **38**, 101547 (2021)
16. Breit, E., Egeland, C., Løberg, I.B.: Cyborg bureaucracy: frontline work in digitalized labor and welfare services. In: Pedersen, J.S., Wilkinson, A. (eds.) Big Data, pp. 149–169. Edward Elgar Publishing (2019)
17. Madsen, C.Ø., Kræmmergaard, P.: The efficiency of freedom: single parents' domestication of mandatory e-government channels. Gov. Inf. Q. **32**, 380–388 (2015)
18. Mullainathan, S., Shafir, E.: Scarcity: Why Having Too Little Means So Much. Penguin Group, Great Britain (2013)
19. Suchy, Y.: Executive functioning: overview, assessment, and research issues for non-neuropsychologists. Ann. Behav. Med. **37**, 106–116 (2009)
20. Walsham, G.: Interpretive case studies in IS research: nature and method. Eur. J. Inf. Syst. **4**, 74–81 (1995)
21. Løberg, I.B.: Efficiency through digitalization? How electronic communication between frontline workers and clients can spur a demand for services. Gov. Inf. Q. **38**, 101551 (2021)
22. Eisenhardt, K.M., Graebner, M.E.: Theory building from cases: opportunities and challenges. Acad. Manag. J. **50**, 25–32 (2007)
23. Döringer, S.: 'The problem-centred expert interview'. Combining qualitative interviewing approaches for investigating implicit expert knowledge. Int. J. Soc. Res. Methodol. **24**, 265–278 (2021)
24. Mayring, P.: Qualitative content analysis. In: A Companion to Qualitative Research, vol. 1, pp. 159–176 (2004)
25. NAV: NAV's horizon scan 2021. In: Welfare, D.o.L.a. (ed.) (2021)
26. Hofmann, S., Madsen, C. Ø., Lindgren, I., Verne, G.: A citizen-centered analysis of what public services are suitable for digital communication channels. In: EGOV-CeDEM-ePart-*, pp. 75–82 (2021)

The Human Touch Meets Digitalization: On Discretion in Digitized Services

Guri B. Verne(✉), Johanne S. Oskarsen, and Tone Bratteteig

Department of Informatics, University of Oslo, Oslo, Norway
{guribv,johansos,tone}@ifi.uio.no

Abstract. This paper investigates how discretion is exercised in today's increasingly digitalized public services. Discretion exercised by street-level bureaucrats is defined as filling in and fitting the rules to the actual circumstances of a case to ensure fair and individual treatment of citizens under all circumstances. We present two empirical cases that illustrates different uses of discretion. We find that discretion is exercised in three stages: in preparing input to a case handling process, during case handling, and in producing the outcome or a decision. Digitalized support for case handlers' discretion is suggested and will be different for these stages.

Keywords: Discretion · Case handling · Digitalization · Plans and situated action

1 Introduction

Public sector case handling is increasingly becoming digitalized, and researchers and practitioners are concerned that the leeway for discretionary assessments is under pressure (Bovens and Zouridis 2002; Busch and Henriksen 2018; Kane and Köhler-Olsen 2018; Hofmann and Busch 2021). By definition, discretion is a concrete interpretation of laws and regulations at the time when they are applied in an actual context (Schartum 2018: 319). The conditions that apply in a particular context or situation cannot be fully predefined, and therefore discretion cannot be substituted by e.g., a list of conditions defined at the time of specifying or programming a digital service (ibid).

When parts of the work processes in the public sector are digitized, more of the "street-level" bureaucrats' encounters with citizens take place through digital self-services. Street-level bureaucrats are often involved in complex case handling aimed at giving individual and fair treatment to people with different experiences, personalities, and circumstances, and can operate with a lot of discretion in decision making. In this paper we explore the nature of discretion aiming at developing an understanding of how and where discretionary assessments take place in digitalized case handling.

Automation and digitalization in different forms have restrained and curbed the use of discretion (Busch and Henriksen 2018; Petersen et al. 2020; Bullock et al. 2020; Holten Møller et al. 2020). According to Bovens and Zouridis (2002) and Howe (in Evans and Harris 2004), much of the power that the street-level bureaucrats previously

R. Krimmer et al. (Eds.): ePart 2022, LNCS 13392, pp. 188–202, 2022.
https://doi.org/10.1007/978-3-031-23213-8_12

had through their use of discretion has shifted to management and the practitioners no longer have the degree of autonomy that Lipsky (2010) suggests. He acknowledges that there are exceptions, such as in the work that cannot be standardized, but which typically requires "in situ judgment" when the street-level bureaucrat is the only resource that can meet the client's needs (Evans and Harris 2004).

The work situations of the street-level bureaucrats are complicated by nature and cannot be reduced to "programmable formats": the situations they find themselves in and the people they meet need a human response, such as compassion and flexibility. The clients believe that the street-level bureaucrats "hold the key to their well-being" and therefore contribute to the legitimacy of the welfare state (Lipsky 2010). However, because of lack of time, information or other resources, the work the street-level bureacrats do is not always possible to perform "according to the highest standards of decision making" (Lipsky 2010, p. xi).

Bovens and Zouridis (2002) distinguish between "screen-level" and "system-level" bureaucracies, where the former illustrates how decision-making is routinized, and data about clients have to be filled in to standardized electronic forms. "System-level" bureaucracies refer to bureaucracies where computer systems automatically make decisions based on data about the citizens. Decisions that could previously be taken by the street-level bureaucrats have thus been pre-made and programmed by someone else, and thus the system developers become key actors. Also, the data that are collected and stored about the person and the circumstances of the case become important. Busch et al. (2018) find that street-level bureaucrats are positive to digitalization supporting their professionalism. In their literature review of digital discretion, Busch and Henriksen (2018) find that digitalization influences street-level bureaucrats' space to exercise discretion to varying degrees. Overall, they conclude that the scope of street-level bureaucracy is decreasing, and that street level bureaucrats to a lesser degree interacts face to face with their clients. However, ICT is not one thing with given consequences for the space for discretionary practices. Busch and Henriksen (2018) use the notion "digital discretion" for "the use of computerized routines and analyses to influence or replace human judgment" (p. 18) and find that it can lead to more fair and equal treatment of the citizens but can also lead to reduced service quality and fail to prevent errors to occur. They call for more empirical studies of the context around the street level bureaucrats' discretionary considerations.

In this paper, we will investigate discretion and how and when it is exercised in public case handling. We will take as a point of departure the ideal that discretion is necessary to fit the rules to the actual situation (where they will be applied) so that fair and equal treatment of citizens in very different situations and circumstances is achieved. The paper is based on our view that discretion that is exercised in a professional way is a necessary part of street level bureaucrats' work and should not be confused with chance or depend on the mood of the bureaucrat.

We draw on welfare research and the field Computer Supported Cooperative Work (CSCW) including our own empirical material from a tax call centre and a welfare administration to describe the practical work of case handlers. Research within CSCW focuses on studying human cooperative work practices as a prerequisite to designing computer systems for supporting the work (Schmidt and Bannon 1992). Studying work practice enables us to recognize what is required from a good practitioner (Schmidt 2011).

Exercising discretion involves interpreting the actual situation where some rules will be applied in a professional way. To analyse and discuss discretion in case handling we use Lucy Suchman's analytical framework (1987, 2007) for human-machine interaction. Suchman suggested a simple form for analyzing interaction where we denote the actions of the human user and the machine including what the user and the machine does that is not visible for the other part. The purpose of this framework is to see if and how humans and computers come out of synch and interaction problems occur. The notions of "plans" and "situated actions" allow us to focus on what is represented digitally in a case handling process and what is not digitally available (yet).

We understand "plans" to stand for predefined rules and the law text but also computer code and internal rules are here seen as plans. Applying rules to the circumstances of a particular situation is seen as "situated action" and we use the notions of "plans" and "situated actions" to analyse and discuss the context and circumstances where discretion is exercised.

In the next section, we describe discretion in case handlers' work and in the laws and regulations. Then we present the notions of "plans" and "situated actions". In section four, we present two empirical cases where discretion is exercised in different ways. In section five, we suggest that discretion can be exercised at different stages in a case handling process, which need different digital support for discretionary assessments. The final section concludes the paper.

2 About Discretion

2.1 Discretion in Case Handling

In a comparative study of governance of the young unemployed in the UK, Germany and Norway, Kane and Köhler-Olsen (2018) analysed the national legislation and job centre conduct in light of the human right to non-discrimination and equality. They interviewed case handlers about their leeway for individual professional discretion in helping young unemployed people improve their opportunities for a job. Understanding the individual and his or her "baggage" is fundamental for finding and suggesting the best activities that will bring them closer to a job. The case handlers exercised discretion in several parts of their follow up of their clients, such as matching jobs and people, tailoring mandatory commitments to each person, and whether and how to sanction non-compliance with compulsory activities as this could ruin what is already achieved (ibid., p. 346). All case handlers "stressed how work-promoting activities and commitments must be tailored to suit individuals with different and complex problem areas" (ibid., p. 347). Discretion can also involve seeking more information about the client's situation that will give a more fine-grained fit to the legislation.

There are boundaries and limits to the space for discretion, mostly set by the law text, and there is also good and weak exercising of discretion. Good use of discretion may be that the case handler asks the client for more information which can support a discretionary decision or asks colleagues if similar cases have occurred previously and if there is precedence from earlier practice. When the case handler has to collect more information and write argumentation for her decision if discretion is exercised, her work becomes more time-consuming (Breit et al. 2020; Løberg 2021). A bureaucrat with a high workload and low competence will make less informed decisions. In some cases, exercising discretion will not be real (Schartum 2018); the citizens' case will be decided according to the rules of law but not necessarily by the purpose of the law. To interpret the law in the particular circumstances of the case is a discretionary decision.

In their study of social workers in Denmark, Petersen et al. (2020) found discretion to be a collaborative effort, where teamwork and interdependency between those involved were part of the decision-making. Discretion was important to help the families in need, and deeply rooted in the law and internal policies used by the social workers. Discretion was not carried out as an individual exercise by a case handler.

In Norway, most citizens' tax return forms are processed automatically except when the Tax Authorities look for tax fraud or money laundering. However, all appeals about the tax assessments are handled manually. Also, other Norwegian digital services exercise automatic decision making with precisely defined decision criteria: decisions can be made almost completely untouched by humans (Mjærum and Cowan 2008; Løvdal 2008). There may be no room for exercising discretion even in cases where an error has been made and is acknowledged: the system is relentless after an automatic decision is made and cannot be undone because the place at school was given to someone else when you erroneously were found to be not qualified (Løvdal 2008). Automation might appear to reduce uncertainty but will not "solve the complexity of situations" (Petersen et al. 2020, p. 329).

Bing (1977) discusses how digital systems introduce a need to set a fixed limit where the rules are unspecific. His early example of how digitalization influences discretion is deciding when the military service is considered completed: could a person skip a week or two at the end of service and still get the service approved? A human would maybe even tolerate three weeks less if the soldier was ill or had good reasons for finishing the service earlier whereas a computer would operate with a preset or calculated date. In this respect, automated digital case handling curbs the space for discretion in some kinds of case handling.

When algorithms replace humans in decision-making at the street-level, challenges related to discretion arise as the algorithms must fill the gap between high-level policy and on-the-ground decisions in novel situations. An algorithm can be designed to learn from historical data. However, it can only be reflexive after a decision has been made and it has been given feedback if the decision was considered incorrect. Humans can use discretion in these cases, as they are able to reflect before the decision is made (Alkhatib and Bernstein 2019).

2.2 Discretion Fills in for Rules

National laws regulate rights and duties for citizens and public agencies. The rule of law secures that the citizens can expect equal treatment under the law, and that juridical decisions that concern them are correct. The rules of law must be applied equally to all citizens in a rational and consequent way (Schartum 1993, p. 31) and "the correct rule must be interpreted and applied correctly" (Boe 1987 in Schartum 1993, our translation).

It also means that the facts or data that a juridical decision is based on are correct and evaluated correctly based on the rules that are applied (Schartum 1993).

The Norwegian law text contains many rules that are imprecise and vague as they contain notions such as "unreasonable" and "special needs" (Schartum 1993, 2018). Such notions will need to be filled with concrete interpretations of what is, e.g., unreasonable at the time when the law is applied in an actual case. Thus, the lawmakers open up for discretionary considerations that may be taken into account in an actual case in the future when the conditions are not known. There are no rules given in the law about what is considered "unreasonable", which can be interpreted differently from case to case. To secure equality for the law in all cases and conditions, discretion exercised by case handlers is necessary for fitting the rule to the conditions of the actual case in the future.

If a law text gives rules and conditions for the interpretation of discretion, it will by definition no longer be discretion (Schartum 2018). Following Bing (1977), who launched the notion of "automation friendly law making", Jansen and Schartum (2008), Schartum (2018) argues that laws made more precise with less openings for discretion will be more suitable for digitalization and thereby for automatic decision making. Specifications and decisions made by the lawmakers will relieve the bureaucracy from exercising discretion at the time of the application of the laws and regulations. However, the laws will be less flexible, less adapted to individual circumstances and more rigid in unforeseen circumstances.

The Norwegian laws are shorter than the laws of many other countries, e.g., US laws are 15 times more voluminous than the Norwegian laws (Nickelsen 2019). Short laws are imprecise and underspecified and contain less text about detailed rules and the conditions they apply to. Some laws open up for more discretion than others, as intended by the law makers. Efficiency is one reason for opening up for discretion in an actual case, as this will allow the lawmaker to not having to specify explicitly all conditions that may be relevant in the future when the law will be applied. Respect for an unknown future is another reason, as an underspecified law that opens up for discretion will be applicable also in unexpected situations and circumstances (e.g., technologies) that are unknown and unforeseen when the law is written. Vague laws create space for discretionary judgments and allow for a delegation of their specification and interpretation to governmental agencies or professions to be carried out in public, in democratic processes (Schartum 2008). Open interpretation of the law increases the public understanding of the law, which contributes to legitimising the law among the law subjects: the citizens (Nickelsen 2019).

3 Situated Action as Analytical Lens

In "Plans and situated action", her seminal work about human machine interactions, Lucy Suchman (1987, 2007) analysed how the actions and decisions of the script-based machine represented the "plans" and what happened in practice as the "situated action". Often the machine in her study came out of sync with the actions and expectations of the humans that operated it. Her analysis explained that this happened because contingencies and circumstances demanded that the user carried out actions without knowing the machine's scripts, and thereby carried out the "wrong" actions. Suchman suggested a simple form for analysing interaction where what the user does that is detected by the machine, e.g. pushing buttons, are "actions available to the machine". The machine's responses that the user can experience are "effects available to the user".

THE USER		THE MACHINE	
I	II	III	IV
Actions not available to the machine	Actions available to the machine	Effects available to the user	Rationale

Fig. 1. Suchman's human-machine interaction framework that focuses on what is available for the other part (Suchman 1987, 2007).

In addition, she suggests to describe more of the actions that the users do around the machine as well as the inner workings of the machine. These notions help us see what the user and the machine does that is not visible, or available, for the other part. In this way, it is easier to see if and how interaction problems occur.

In this paper, we see the "plan" as the predefined laws, rules, computer programs and internal instructions and regulations that are involved in a street-level bureaucrat's decisions and the "situated actions" as what he or she does in an actual case handling process. A juridical perspective on discretion emphasizes that laws always must be fitted to the actual case, the situation, and that it is not possible to describe all relevant situational characteristics in advance. Discretion can be defined as the exercise of interpretation and judgement within the space for action that the rules open for.

Automation and ICT can be understood as "plans" as they represent and implement rules that are decided on in advance. As plans they are resources for the situated action in Suchman´s analysis (2007), and discretion is what helps adapting the plan to the actual circumstances when the plan is carried out.

Interpretation of the law in a particular case therefore involves the law as well as internal rules and regulations and previous interpretations and applications of these. Exercising discretion requires some degree of expertise concerning both the rules and the interpretation of the situation. Experts exercise their knowledge and expertise in the situation (Goodwin 1994; van Hout et al. 2015; Orr 1996). Novices tend to follow rules while experts act based on their experiences and intuition (Dreyfus and Dreyfus 1980). Just following rules means not being better to assess a case than a novice.

4 Empirical Examples of Discretion in Tax and Welfare

In this section, we will present examples from two cases based on two of the authors' fieldwork. The examples illustrate how discretion is exercised at "street level". The first two examples are from of a study on how the automation of taxes influences the work the citizens do when they do their taxes (Verne 2015). The next two examples are from a study of how digitalization influences the work of the welfare advisors in a welfare agency (Oskarsen 2020).

4.1 Call Advisors in the Tax Call Centre

We start with a couple of examples from the fieldwork of one of the authors Verne (2015), focusing on the practices around a semi-automated tax service. The fieldwork involved several hours of listening in to 474 telephone calls to the Tax Information Call Centre (TICC). We took notes, carried out observations and 15 in-depth qualitative interviews of call advisors and others in the tax administration that gave a direct insight into some problems that the citizens struggled with. The material was analysed inductively and is documented in (Verne 2015). Further examples are described in (Bratteteig and Verne 2012; Verne and Bratteteig 2016). For this paper, to illustrate how discretion is exercised we have selected two calls to the tax call centre that ask for relatively similar tax advice but receive very different responses from the two advisors.

Example 1: The caller presents herself as the mother of a young man 18 years old who receives disability pension. The mother is his trustee. Her request concerns the tax card that specifies the advance tax deduction percentage from his income. It is preset to 2% of his income, and the mother asks the advisor to increase it, just to be sure, so that he will be protected against future extra tax bills if his income increases during the year. The mother hopes that the son "will get himself a little job" this year and she does not expect that he in this case can follow up the tax deductions himself. The call advisor Berit asks for a confirmation that he has a disability pension, and then replies that the mother must send a written request to the welfare administration that pays the pension and deducts the advance tax before paying him the pension. Berit replies that to issue a new tax card she "has to correct the figures" and she refuses to do so as long as his income has not changed. The mother reformulates her request and receives the same answer. Berit gives some practical advice on how the mother can change the income figures herself in the online self-service, and the mother close the call with an angry "Gosh. Oh my God!".

An employer is required by law to increase the advance tax percentage if an employee asks. This information is not easily available on the tax webpages, and Berit does not mention it. The internal routines state that the advisor shall encourage the callers to use the digital self-services, and Berit followed the routines even though this caller asked explicitly for help for her disabled son. Some of the advisors think that "somebody has to stand up for the rules" (Verne 2015). Following the rules in a literal sense leaves little room for discretion, as in this example where some elements of the actual situation could have been given weight to a discretionary response from the advisor to help the mother help her disabled son.

Example 2. A young woman calls. She starts the conversation by saying that she thinks her advance tax deduction is too low. She is only deducted 16% advance tax from her income, and she wants to avoid being billed for outstanding tax later. The advisor asks promptly if she is a single parent, which she denies. The advisor proceeds by asking her about her salary and offers to increase the percentage registered in the database for advance tax. She does not mention that the caller can do so herself on the tax web pages. After calculating a little by hand, the advisor suggests 20% monthly deduction, which they agree on. Then the advisor changes the data about income in the advance tax database and issues a new tax card with 20% deduction.

In the second example the call advisor changed the income figures (that the first advisor refused to do) and helped the caller directly, even though the need for some extra service to the disabled son seemed stronger in the first example.

We have selected these examples as they illustrate how discretion can be exercised and the meaning it will have for the citizen that are influenced by it. Advisors at the call centre do not carry out case handling nor make decisions in cases. This fieldwork also includes examples where the call advisor enters data for a caller that is in a life crisis and where the advisor believes that the caller will lose money if he does not succeed in changing the online income data without help. Discretion is in these examples exercised as in how the advisor meets the caller, how s/he explains the rules and the online service, and how willing s/he is to go some extra lengths to help the caller when it may be in the caller's best interest, even when it implies overriding some internal routines (Verne 2015).

4.2 Advisors in the Welfare Administration

The next couple of examples come from the fieldwork of a second author (Oskarsen 2020). This fieldwork is from the welfare administration and is based on in-depth interviews and observations of welfare advisors who work directly with clients. We use the term advisor for the welfare personnel with direct contact with the client. The advisor's role is to help the client find or return to work or apply for financial benefits if working is not possible due to, e.g., health-related issues. A client should always have their work ability assessed by the advisor. The work ability assessment is a document describing the client's case based on the available documentation, such as medical certificates, reports from courses, education, etc. The case handlers do not meet the clients but make the final decision in a case based on the available documentation, such as the work ability assessment document and the relevant laws and regulations.

Example 1. Siri is an advisor who works with young adults receiving work assessment allowance, a temporary financial benefit. One of her clients is a young woman who recently has undergone cancer treatment and is considered to be recovered. However, she is suffering from fatigue and is not able to continue working. Siri considers that this client is entitled to disability benefits, a permanent financial benefit, so that she can focus on full recovery from her fatigue so that she will be able to work in the future. However, the client's application for disability benefits was rejected by the case handler, arguing that the client was able to do some work-related activities, such as trying out more work arrangements. Siri disagreed and set out to look for a way of presenting the facts in the case so that the description better met the reasons for the rejection. She obtained additional information, such as the hospital's medical history, which had not been part of the documentation that she based the work ability assessment on and spent months presenting a better argumentation in her client's favor. In the end, her client was granted disability benefits.

Example 2. Peter is an experienced advisor who explains that writing a good work assessment document takes time. It often starts long before any actual document is opened. The client, advisor, and doctor need to have a common understanding of the client's situation and agree on what is best for the client. Here we look at a client's case that is difficult to present in a good way, as he lacks a clear diagnosis. However, as Peter has met with the client multiple times over the years and followed his failing health and his attempts to try out several different types of work. Peter is convinced that the right decision for the client is to be granted disability benefits. He starts writing a draft that contains the main points, and goes back to the text many times and changes the formulations, and tweaks the arguments. Peter's loyalty lies with the client, and he wants to make the argumentation foolproof according to the laws and regulations so that the case handler cannot reject the assessment.

In both examples, discretion is involved in how the advisors choose to follow up the client and spend the effort necessary to document important elements of the case and write a good work assessment that is not rejected internally in the welfare administration. Citizens come to street-level bureaucracies as unique individuals. For them to receive treatment or help from the bureaucracy, the complex citizen must transform into a client that is manageable for the bureaucratic system. People are therefore represented as one of a small number of categories defined by the bureaucracy and associated with a rule or civic rights. These client characteristics often do not exist outside the process that gives rise to them; the social process it is to make a human being into a client (Lipsky 1980). Clients are both consumers of the bureaucracy's output or services and the "raw material" that the bureaucracy processes (Prottas 1978). The categorization of the welfare client is not given, and the client's problem and its solution are created together (Skaarup 2021).

In the examples above, we have illustrated that discretion can be exercised in many forms and stages in a case handling process, from the way a caller is first met and helped to proceed with their own affairs to how the advisor works iteratively with providing and documenting relevant information and writing an assessment text.

There are different degrees of complexity and possible consequences of the decisions made in the examples above; the call advisors make decisions on how to help the caller but the case handlers make decisions about economic and health related support. Discretion in the tax call centre examples can be understood as taking place before the system-level automated case handling where the data in the tax card will be digitally produced. Tax-relevant cases are often about calculations, although the citizens' life situation will be of relevance to which calculation rules will be applied. In the welfare area the clients' whole situation needs to be understood and interpreted against the rules, often with a look to similar cases and how they were interpreted and presented for a benefit being granted. Discretionary assessments are made more or less explicit and in several rounds. The advisor's willingness to work with an assessment is an important aspect of discretion although it gives no guarantee for the result.

5 A Conceptual Model for Stages of Discretion

Based on our investigation of discretion based on previous research and own empirical material, we see that discretion can play different roles in different decision processes. Discretion can be exercised in different stages in a digitalized case handling process:

the data that are the input to computer algorithms can be based on discretionary considerations, how the context where laws and rules are applied in a case can be subject to discretionary considerations, and the decision itself can be a direct result of discretion. We see this as discretion as input to a decision process, exercised during case handling, and as the basis for the output of a decision process, see Fig. 2.

Discretion for Input: Data based on discretionary assessment may be entered into databases and become the basis for computer-based decisions. Discretion may be exercised when preparing the data that will form the basis of a computer calculation. An example is the Norwegian tax rules stating that when (parts of) a citizen's income cannot be documented, discretion-based figures may be estimated and used for income tax calculations, such as for e.g., waiters or taxi drivers. In this case, a figure based on discretion is input into the computer calculations for the income tax for the waiter or taxi driver.

Discretion During Case Handling: Discretion may be exercised as in the welfare cases above in how the advisors work with preparing the assessment documentation. The data and the rules need to be considered together, and the advisor can choose to collect more documentation that can strengthen the client's case. A new round of case handling will be based on the advisor's belief that the case can benefit from a better assessment – again a discretionary assessment.

Looking for precedence from similar cases is an important element in case handling for coming to a decision or assessment in line with the law and previous practice (Blomberg et al. 1996). What will count as similar enough will often be based on judgment, creativity, and discretion. Several elements in one case may be similar to another case, but some elements may be considered so different that the case as a whole is not seen as similar. Such considerations will often be part of juridical assessments in case handling. Here discretion is exercised when comparing the case in hand with candidates for similar cases. It is also interesting to discuss discretion as decisions relating to the purpose rather than the letter of the law, i.e., to judgement that goes beyond the space for action that the rules open for.

Discretionary Decision/Result: Discretion may also be exercised when coming to a decision, as the case described by Schartum (2018) about how a woman on social support one year earned five dollars more than the year before and this made her no longer qualified for social support: she literally lost several thousands of dollars in exchange of five dollars more in income. In this case, the unfairness of this consequence was considered serious, and she discretionally was granted her support although she earned more than the rule literally stated. In some cases, a decision is taken first, and the rules are applied afterwards: retro-fitting the rules to the preferred decision. An example of retrofitting is a woman negotiating her diagnosis with her psychoanalyst about to achieve both the insurance benefits of having a diagnosis and avoid social stigma from having a psychiatric diagnosis. They agreed on obsessive-compulsive disorder as a compromise (Bowker and Star 1999: 47). Here the patient/client had control of the category she was to be placed in, where she considered the consequences for this category as part of

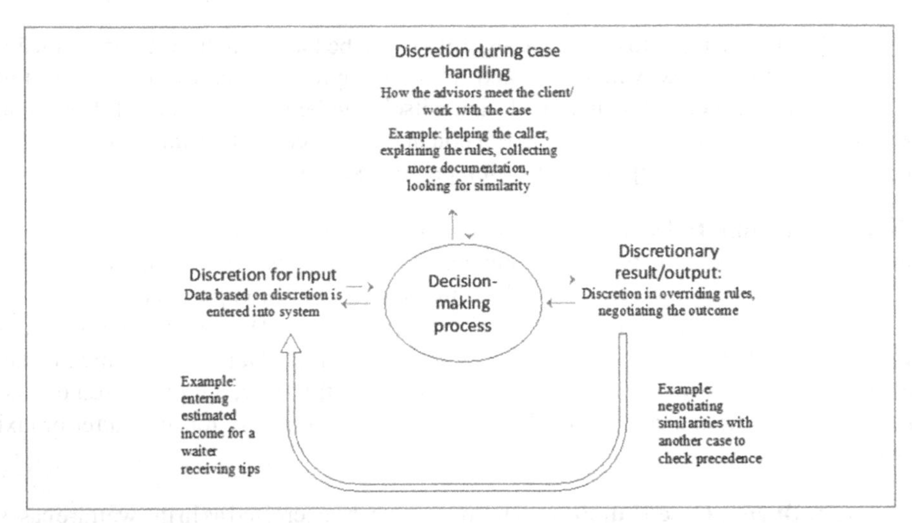

Fig. 2. Discretion may be exercised when preparing a case handling process, during case handling or as the result from case handling and can take place in several rounds.

deliberating her decision. In such cases the purpose of the law is used to identify which law text or regulations are utilized for the desired outcome.

An important aspect of discretion is the advisor's willingness to follow up and work to achieve the best decision for the citizen/client. Discretion can take place in several rounds, and after a rejection a discretionary assessment may lead the advisor to try a new round where more and better documentation is collected for the case. An advisor's previous experience gives a large "repertoire" for considering similarity with other cases and improve the chances for a new round to succeed.

6 Digital Discretion – A Contradiction in Terms?

Discretion is exercised when the rules are applied and interpreted in actual circumstances and is necessary in complex cases (Busch et al. 2018; Kane and Köhler-Olsen 2018; Petersen et al. 2020). In Suchman's (2007) terms, discretion will always as a minimum be exercised when adapting the plan to the situated action, in case handling terms adapting the laws to the actual situation of the case. Researchers disagree whether digitalization constrain or support discretion (Jorna and Wagenaar 2007; Buffat 2015; Petersen et al. 2020). Schartum (2018) states explicitly that if the rules specify the basis for a decision in detail, it will not be a discretionary assessment. Schartum (ibid.) indicates that if lawmakers add more and more detailed specifications to the law it will still leave discretion outside of the specifications of the law. There will always be some leeway in how a situation is understood and found to match laws and regulations.

Digital discretion is in this perspective a contradiction in terms, but in line with Møller et al. (2020) we argue that discretionary assessments can be digitally supported and presented instead of being digitalized away. Our concern in the following is how case handlers' exercising of discretion can be digitally supported. Digitalized support for discretion will be different for the different stages in Fig. 2. For supporting discretion in the input stages, an estimated income figure based on previously processed figures can be presented by the system. During case handling, a system may suggest similar cases based on a rich set of criteria, like text searches on the Internet but augmented with more than text search, for example synonyms, related concepts, contrasting concepts etc. Digitalized support for the decision or result stage will be different, as it concerns overriding specific criteria in special situations to make a decision. Experts will know what to look for, and not necessarily bother with all the other criteria (see e.g., Dreyfus and Dreyfus 1980). Easy access to overruling automated system decision will be support for discretion in this stage.

Schartum defines discretion as the interpretation and judgement that computers cannot do (1993, 2018), because they follow rules. In Suchman's terminology, this means that discretion can never be specified and calculated in the plan. However, we will argue that a digital system building on historical data can make decisions that include discretion because the discretion is built into the data from previous decisions. We would not say that the digital system exercise discretion but that the data it builds on includes previous discretionary decisions.

Such digitalized discretion support has its limits as it builds on historical data with historically decided parameters. Data are not neutral, and the circumstances of their production and use is important for their interpretation (Gitelman 2013). It is possible that a case will have to emphasize a new parameter or shift the weighting of parameters in a new way compared to all historic cases. The adjustment to the situation at hand may be different from most other historic cases. The extreme case is when a shift occurs within the actual domain. An example from medicine illustrates this point: the AIDS diagnosis took years to be recognized since the symptoms of the disease were not recorded in a way that made comparison between similar cases possible. Shifts like discovering a new diagnosis like AIDS start as strange and unusual situations and it may take years before the characterizing parameters are included in the historic data as the normal basis for decision-making (Bowker and Star 1999). The discretion exercised when seeing new characteristic in a situation and recognizing their significance requires a human. However, accessing and processing of possible parameters can be done efficiently by a computer.

Machine learning (ML) is a technology for searching for patterns in large data sets and through statistical calculations produce decisions based on approximations to the current case from historical data (Goodfellow et al. 2016). We can consider ML-based systems to process and present data from historical cases to support human discretionary practices instead of automating away the room for exercising discretion. However, in image recognition, where deep learning has provided some early successes, it has also been found to give unstable results: just adding a few pixels that can not be seen with a human eye in an image changed the machine interpretation considerably, and to an opposite conclusion (Nguyen et al. 2015; Antun et al. 2019).

What will such recently discovered instabilities mean for the decisions based on complex, heterogenous but incomplete data collected and stored in public services? We argue that it will be important for citizens to have the possibility to critique case handling made on the basis of Machine Learning (Coombs et al. 2021). Transparency of the decisions made by ML-based systems will be important for trusting the decisions and criticizing them (Jobin et al. 2019; Dignum 2018).

7 Conclusions and Future Research

Laws and rules are often underspecified with respect to the circumstances where they will be applied. Street-level bureaucrats' discretion is necessary to fill in for the rules and fit them to the actual situation. Digitalization of case handling has diminished or moved the space for discretion as previously discretionary decisions are now built into computer algorithms where the criteria are specified in detail. However, we do not see such algorithmic decisions as discretionary as they are made into rules. There is room for both rules (plans) and discretion (situated actions), as the rules are a resource for exercising discretion.

Based on examples from research and the authors' fieldwork, we suggest that human discretion is exercised in three stages in a possible case handling process:

Discretion for input to a case handling process.
Discretion during case handling.
Discretion for the result/output from a process.

Discretionary assessments can take place in several rounds. They influence how a citizen is first met by an advisor or case handler, how they understand the case and find similarities with other cases, and how much effort they put into the work of preparing documentation and writing an assessment. For the discretion to be real a case handler need to have the time for more rounds.

We are convinced that computer systems can support discretion in public services. The stages of discretion that we suggest here can help us discuss how computers can support human discretion. Designing support for discretionary judgments will be different for these stages. Digital support for human discretion can be a resource for a better fit between the rules and the circumstances of the individual cases, which is important for fair and equal treatment by the law. Future empirical research can bring more knowledge about how and when discretion is carried out in digitalized public case handling.

We suggest that it is possible to experience, or simulate, digital discretion when a computer system bases its decisions on historical digital data that include decisions based on discretion. However, this discretion is interwoven with the data (and the statistics) and is not open to explanations or challenges like human discretion is and must be.

Future research can study if digital data representations and ML approximations can become good enough to replace the analogue or human original. In some cases, they can, e.g., in photography and medicine, but in other more open and complex cases where different types of characteristics are weighted against each other they may not. It will be an important topic for future research to study if we are capable of distinguishing between the cases where the approximations are good enough from the cases where they are not.

How an advisor understands the situation and how much effort they will put into making a case so that a fair decision is made will always be a matter of discretion. Citizens that experience that their case is handled fairly is the basis for a legitimate state.

References

Alkhatib, A., Bernstein, M.: Street–level algorithms: a theory at the gaps between policy and decisions. Paper Presented at the Proceedings of the 2019 CHI Conference on Human Factors in Computing Systems, Glasgow, Scotland, UK (2019)

Bing, J.: Automatiseringsvennlig lovgivning (automation friendly law making). Tidsskrift for rettsvitenskap 195–229 (1977)

Bovens, M., Zouridis, S.: From street-level to system-level bureaucracies: how information and communication technology is transforming administrative discretion and constitutional control. Public Adm. Rev. **62**(2), 174–184 (2002)

Bowker, G.C., Star, S.L.: Sorting Things Out: Classification and Its Consequences. Massachussets Institute of Technology, Cambridge (1999)

Bratteteig, T., Verne, G.B.: Conditions for autonomy in the information society: disentangling as a public service. Scand. J. Inf. Syst. **24**(2), 51–78 (2012). Article 3

Breit, E., Egeland, C., Løberg, I.B., Røhnebæk, M.T.: Digital coping: how frontline workers cope with digital service encounters. Soc. Policy Adm. **55**(5), 833–847 (2020)

Busch, P.A., Henriksen, H.Z.: Digital discretion: a systematic literature review of ICT and street-level discretion. Inf. Polity **23**(1), 3–28 (2018)

Busch, P., Henriksen, H.Z., Sæbø, Ø.: Opportunities and challenges of digitized discretionary practices: a public service worker perspective. Gov. Inf. Q. **35**(4), 547–556 (2018)

Buffat, A.: Street-level bureaucracy and E-government. Public Manag. Rev. **17**(1), 149–161 (2015)

Bullock, J.B., Young, M.M., Wang, Y.: Artificial intelligence, bureaucratic form, and discretion in public service. Inf. Polity **25**, 491–506 (2020)

Coombs, C., et al.: What is it about humanity that we can't give away to intelligent machines? A European perspective. Int. J. Inf. Manag. **58**, 102311 (2021)

Dignum, V.: Ethics in artificial intelligence: introduction to the special issue. Ethics Inf. Technol. **20**(1), 1–3 (2018). https://doi.org/10.1007/s10676-018-9450-z

Dreyfus, S.E., Dreyfus, H.L.: A five-stage model of the mental activities involved in directed skill acquisition. Operations Research Center, University of California, Berkeley (1980)

Evans, T., Harris, J.: Street-level bureaucracy, social work and the (exaggerated) death of discretion. Br. J. Soc. Work. **34**, 871–895 (2004)

Gitelman, L. (ed.): "Raw Data" Is An Oxymoron. MIT Press, Cambridge (2013)

Goodwin, C.: Professional vision. Am. Anthropol. **96**(3), 606–633 (1994)

Goodfellow, I., Bengio, Y., Courville, A.: Deep Learning. MIT Press, Cambridge (2016)

Hoffmann, S., Busch, P.A.: Citizens' perception of digital discretion. In: SWEG 2021, Copenhagen (2021)

van Hout, A., Pols, J., Willems, D.: Shining trinkets and unkempt gardens: on the materiality of care. Sociol. Health Illn. **37**(8), 1206–1217 (2015)

Jansen, A., Schartum, D. (eds.) Elektronisk forvaltning på norsk. Statlig og kommunal bruk av IKT. Fagbokforlaget Vigmostad & Bjørke, Bergen (2008)

Jobin, A., Ienca, M., Vayena, E.: The global landscape of AI ethics guidelines. Nat. Mach. Intell. **1**(9), 389–399 (2019)

Jorna, F., Wagenaar, P.: The 'iron cage' strengthened? Discretion and digital discipline. Public Adm. **85**(1), 189–214 (2007)

Kane, A.A., Köhler-Olsen, J.: Governance of the young unemployed – a comparative study of the United Kingdom, Germany and Norway. Eur. J. Comp. Law Gov. **5**(4), 317–377 (2018). https://doi.org/10.1163/22134514-00504002

Lipsky, M.: Street-Level Bureaucracy: Dilemmas of the Individual in Public Services. Russel Sage Foundation, New York (2010)

Løberg, I.B.: Efficiency through digitalization? How electronic communication between frontline workers and clients can spur a demand for services. Gov. Inf. Q. **38**(2), 101551 (2021)

Løvdal, E.: Samordna opptak: norge samlet til ett utdanningsrike? In: Jansen, A., Schartum, D. (eds.) Elektronisk forvaltning på norsk. Statlig og kommunal bruk av IKT. Fagbokforlaget Vigmostad & Bjørke, Bergen (2008)

Holten Møller, N., Shklovski, I., Hildebrandt, T.T.: Shifting concepts of value: designing algorithmic decision-support systems for public services. In: Proceedings of the 11th Nordic Conference on Human-Computer Interaction: Shaping Experiences, Shaping Society, pp. 1–12. Association for Computing Machinery, New York (2020). Article 70. https://doi.org/10.1145/3419249.3420149

Nguyen, A., Yosinski, J., Clune, J.: Deep neural networks are easily fooled: high confidence predictions for unrecognizable images. Paper Presented at the Computer Vision and Pattern Recognition (2015)

Nickelsen, T.: Den nordiske lovgivningsmodellen: Norge har verdens korteste lover. Apollon Res. Mag. **2**(29), 30–34 (2019)

Orr, J.E.: Talking About Machines: An Ethnography of a Modern Job. ILR Press, Ithaca (1996)

Oskarsen, J.S.: Closing the loopholes: categorizing clients to fit the bureaucratic welfare system. In: ACHI 2020 The Thirteenth International Conference on Advances in Computer-Human Interactions, Valencia, Spain, 21–25 November 2020 (2020)

Petersen, A.C.M., Christensen, L.R., Hildebrandt, T.T.: The role of discretion in the age of automation. Comput. Support. Coop. Work (CSCW) **29**(3), 303–333 (2020). https://doi.org/10.1007/s10606-020-09371-3

Prottas, J.M.: The power of the street-level bureaucrat in public service bureaucracies. Urban Aff. Q. **13**(3), 285–312 (1978)

Schartum, D.W.: Rettssikkerhet og systemutvikling i offentlig forvaltning. Universitetsforlaget, Oslo (1993)

Schartum, D.W.: Digitalisering av offentlig forvaltning - Fra lovtekst til programkode. Fagbokforlaget, Bergen (2018)

Schmidt, K.: The concept of 'work' in CSCW. Comput. Support. Coop. Work **20**, 341–401 (2011). https://doi.org/10.1007/s10606-011-9146-y

Schmidt, K., Bannon, L.: Taking CSCW seriously. Comput. Support. Coop. Work **1**(1), 7–40 (1992). https://doi.org/10.1007/BF00752449

Skaarup, S.: Beyond substantive goals – a framework for understanding citizens need and goals in bureaucratic encounters. In: Scholl, H.J., Gil-Garcia, J.R., Janssen, M., Kalampokis, E., Lindgren, I., Rodríguez Bolívar, M.P. (eds.) EGOV 2021. LNCS, vol. 12850, pp. 86–102. Springer, Cham (2021). https://doi.org/10.1007/978-3-030-84789-0_7

Suchman, L.A.: Plans and Situated Actions: The Problem of Human-Machine Communication. Cambridge University Press, Cambridge (1987)

Suchman, L.: Human-Machine Reconfigurations: Plans and Situated Actions. Cambridge University Press, New York (2007)

Verne, G.: The winners are those who have used the old paper form. On citizens and automated public services. Ph.D. thesis, Department of Informatics, University of Oslo (2015)

Verne, G., Bratteteig, T.: Do-it-yourself services and work-like chores: on civic duties and digital public services. Pers. Ubiquit. Comput. **20**, 517–532 (2016). https://doi.org/10.1007/s00779-016-0936-6

Author Index

Printed in the United States
by Baker & Taylor Publisher Services